Altruism is innate but it’s not instinctual.
Everybody’s wired for it
but a switch has to be flipped.

~ *David Rakoff*

# Other books by Graham Mole

A Multitude of Fins (Bennison Kearny UK 2014)

# Paper Chain

by

**Graham Mole**

Silver Bow Publishing

720 – Sixth Street, Box # 5
New Westminster, BC
V3C 3C5 CANADA

Title: Paper Chain
Author: Graham Mole
Publisher: Silver Bow Publishing
Cover Photo: Buntzen Lake, BC CANADA
courtesy of Charlotte Hill
Cover Design: Candice James



www.silverbowpublishing.com
info@silverbowpublishing.com

Library and Archives Canada Cataloguing in Publication

Title: Paper chain / by Graham Mole
Names: Mole, Graham (Journalist), author.
Description: Previously published by YouWriteOn in 2011.
Identifiers: Canadiana (print) 20190100907 | Canadiana (ebook) 20190100915 | ISBN 9781774030318 (softcover) | ISBN 9781774030349 (HTML)
Classification: LCC PR6063.O429 P37 2019 | DDC 823./914—dc23

**Dedication**

To my wife Annie, not just for her meticulous proofreading but her tolerance and patience on the difficult days that turn me grumpy; and to an endless number of foresters, academics and others involved in caring for the planet's trees. They truly are lifesavers and we should all listen to their advice.

## CHAPTER ONE

So, after all the fuss, the rogue space lab had landed. Nobody was killed, the insurance companies won their bets and the reporters, a touch sadly, called off their siege of the NASA base. President Lean, a week now in the White House, allowed himself a half grin. Some of the stories had been bizarre mirrors of a national talent for over-reaction. But there had been serious concern too -- and he'd shared it.

The daily digest of press coverage was on his desk. He picked it up, walked to the window and glanced through it there. The space lab, long thought of as redundant, had been sent up by his predecessor, "Hard-line Hank" Daley. After a year, its work had been finished. Then, inexplicably, its orbit had suddenly decayed. A frail rod, connecting its two 20-ton spheres had snapped, making it impossible for NASA to control its re-entry. It was that lack of control that had made the headlines.

When it was launched the plan had been to retrieve the spacelab with a shuttle. For that reason, it had been fitted with heat shields. Gleefully, almost, the media had speculated where the 20-ton twins would land, what damage would be caused and who, if anyone, would be hit. They'd had a field day.

Now that story at least was over. One had hit Indonesia; the other plunged into a national park in Oregon. Nobody dead, nobody hurt. Lean doubted it would make a line in the next editions. He'd survived his first crisis.

Though Daley had put up the spacelab in the first place, Lean, in opposition, had backed the idea on condition there was room for scientific experiments by other nations.

In all the excitement of moving into the White House he'd almost forgotten that until reminded by his deputy Joe Simon.

Lean smiled to himself. Trust Joe to remember. Perhaps it was understandable. Joe had spent much of his time as an ecologist with the

Earthwatch outfit. It had slaughtered Daley's vote in California and helped enormously with the liberal and green votes. *"Green with Lean"* and *"Clean with Lean"* had both been good slogans.

Joe, being Jewish, had been a gamble but, in the end, it had worked. He must call him, mused Lean. He could pull his leg about the spacelab landing in Joe's home state of Oregon.

Lean felt good now. He was in charge. He put up his feet on the Oval office desk and called Joe's private number. Joe took the call, recognised Lean's voice and gestured to his guest radio reporter, Bob Powell, asking him to hang on for a couple of minutes. Powell moved away to the window out of earshot. But, to Lean, Joe still sounded distracted. He explained he had someone with him.

"Don't worry," said Lean "it's not important. It can wait."

What is it?" asked Joe.

"I just wondered if you ever did get the details of what scientific experiment went on at that spacelab. I guess it doesn't matter too much now, but did you find out?"

"Yes -- I've been trying to get hold of you," said Joe, curtly.

"Trouble?" guessed Lean.

Joe glanced over his shoulder. Powell was still by the window, but he dropped his voice lower anyhow. "A can of worms. I need to talk to you."

"Soon as you like," said Lean, recognising the tone.

"With you in ten minutes" said Joe. He returned his attention to his guest. They each had a lot to gain from Powell's interview.

For Joe, first Jewish Vice-President, it was the chance to make a few points, even to fly a few kites. He could float an idea and gauge the public reaction before putting it into the government machine for fine tuning.

For Powell, it was prestige for his newly acquired show, a mix of news and gossip -- The Folk on the Hill. For both it was part of the dialogue of power, an arrangement as old as politics itself, a trade with delicate counters and balances. Powell was intrigued by the call. It had to be the President. But about what?

Joe settled down again for the interview and, to save time, Powell recorded his introduction on the spot. As he went back to the machine he realised he'd left it running. But there was still plenty of tape left. He did his introduction… the States knew about Joe Simon now, but

Powell filled in more details than usual, knowing the show was likely to be re-broadcast to the world on Voice of America. The extra fee was always handy.

"Ask anybody in town what he knows about Joe Simon and it's my guess you'll only get two answers -- he's Jewish and he's Vice President. Being either of those things can be difficult. Both together could give a guy ulcers. But it's my bet that Joe Simon just won't have that trouble. For one thing he goes home nights -- not always the case on The Hill -- and for another, his wife Ruth cooks great food -- and I can vouch for that." It's part of his religion; looking after the family. That family's larger now and while it might have enough to eat he does worry about the air we all breathe. Joe comes from Oregon where they've been breeding ecology politicians since the days of Governor McCall. That influence made Joe quit his job as a lawyer with a multi-national and get into politics. Now his nickname is *'the President's conscience'*. But maybe there's another one – *'the non-average Joe'*." Powell flicked off the machine, grinned again. "Just think what I could have said."

"Sure" said Joe "but would you believe I'm sick of crucifixion jokes."

Powell flicked the machine on again. The interview started, Powell probing for Joe's reaction to his first week in the White House. Joe replied, "You know Bob politics is a crazy business. Where else would you spend four years trying to get hold of a business where the books are locked away from you? That's just not Jewish. So, this week I've been looking through the books."

"Are they bad?"

"Financially they're OK, but there's a lot of decisions -- big ones -- to be made fast. Firstly, do we keep pouring money into the National Car Corporation? Why put our money into a device that mostly just pollutes our atmosphere? Maybe we should have a really serious crack at an electric car." He paused, gulping slightly. "And there's the Middle East. Sounds odd for a Jew maybe, but don't we have to be supportive to the PLO in a non-token way instead of just hoping they'll go away?"

Powell interrupted: "How about your main interest, the environment? You were quoted the other day as saying Americans were sick."

Joe paused. His politician's instinct flashed "red alert". He could see those headlines again. "Put it this way" he said "I think it's time America took on its role as world leader in a different way. But what we actually need to do would probably cause a revolution. We need to use about half the current amount of gasoline, less paper, re-cycle some...."

"Why paper?"

"Trees."

"Yes, but...."

Joe got into full flow. "The world's forests are going fast. When they go, we go." The phone trilled. Joe cursed and answered. Suddenly he was subdued. He waved at Powell to switch off again. This time Powell did so.

"Yes. Sure. On my way in just a minute."

Powell's ears were at full stretch; certain it was the President calling. Studiously casual he asked, "Something urgent?"

"John Lean isn't an urgent kind of guy. You know that. But he did say five minutes. Let's go quickly."

In five minutes, flat Simon rattled through his main concerns, headline fashion. He told him about how his two boys had teased him about White House waste of paper. Now, paper would all be re-cycled, for the first time. He filled him in on his plan for a 'good ideas' bank and how he wanted less government, not more, encouraging families where applicable to take over welfare functions left previously to the state.

Powell thanked him and, conscious of Joe's obvious haste, signed off. They thanked each other and Powell left. Joe rummaged through the mess of papers on his desk and withdrew a single sheet he'd hidden before Powell arrived.

***

Joe scuttled into Lean's office. The President was back at the window, his right hand tapping his thigh. Joe put the piece of paper on the empty desk. "You might need to sit down."

Lean read it. His left hand slowly stroked his jaw, as if checking his shave; a slow, thoughtful looking gesture. Simon had seen it before, during the campaign. With Lean it spelt tension as he recognised trouble. Beneath the NASA logo the words were stark:

*"Each sphere of the spacelab contained cultures which were part of a Japanese scientific experiment. The cultures were of pine wilt nematodes kept under control by a quantity of new insecticide. The result of the experiment is not known since control of the satellite -- on the direct orders of the President -- was passed to the Defense Department."*

There was a footnote:

*"Pine wilt nematodes are, in lay terms, worms which penetrate trees causing their death. They have killed large numbers of trees in Japan and some in North America. The effect is similar to Dutch Elm disease. No control yet exists."*

Lean looked up at Simon. "You're right -- a can of worms. What does it mean? What do we do next? Presumably they came back in the capsules. But are they alive or dead? There's a million questions."

Joe nodded.

Lean's hand went back to his jaw. "And why would the Pentagon take over control of a satellite full of worms."

Joe shrugged and spread his hands wide. "Who knows with Daley? I hate to say it, but I've got a feeling there's a lot more to it."

"Like what for God's sake?"

"John, I just don't know. I just have a nasty feeling about it."

Lean got up angrily and paced the room, hand again tapping his thigh.

"You don't think Daley was getting us into germ warfare do you?" Joe shuffled himself on to Lean's desk, legs not quite touching the ground.

"John listen. There's a couple of things we have to do first. First we find out if there's a problem. Right?" Lean's pacing slowed slightly. He nodded. Joe went on. "We need to know if the worms -- the PWN -- are alive. We've got to find out what else there was on board. It can't have been just worms. In other words, you've got to get the information out of the Defense Department. But quietly. We can't have any of this go public."

"You're damned right it has to be quiet. I'm not taking blame for anything Daley did."

"OK" said Simon. He continued "I'll make a few quiet calls here and there and find out some more about these PWN worms. And John. Don't worry. It hasn't happened yet. I hope."

Lean laughed as Joe retrieved his piece of paper and bustled from the office. He went back to the window, hand again at his jaw. 'What a mess' he thought 'a week in office and this happens. I can see the headlines … *bugs from space*. Try to explain that away to the Russians and the rest of the world.'

***

Back in his office Joe Simon made a decision of his own. He'd recruit help to assess the risk from PWN. Two minutes later he was talking to John Foster, head of the UN's Food and Agriculture Organisation, asking him to nominate a helper. Foster's recommendation was clear. Dick Walton. A bit young, maybe mid-thirties, but a rising star and he had published some good stuff on deforestation.

***

At the radio station Powell started to edit his tape. He spooled to the start to find his introduction, went too far and heard Simon talking to the President. Suddenly Simon's voice dropped to a whisper. Powell strained his ears. He could just make out the words "A can of worms."

What on earth, he wondered, was that all about? And why was it so urgent? Was it to do with the satellite? And, if so, what did it have to do with it?

## CHAPTER TWO

Walton woke sweating. The mid-afternoon heat of the jungle was building.

Once again the monsoons hadn't happened. The rain that should have beaten its deafening tattoo on the tin roof had taken another day off. He sat up, rubbed his face in a towel, swung his legs to the edge of the camp bed -- and saw the message.

It was in two parts. One part was official. "End mission. Return soonest. Urgent. Foster." His boss, Foster of the FAO, part of the UN. The other part of the message was from Grace, Foster's secretary: "Miss you. Rush home soon. Love you." Both messages had been relayed from Washington over the field phone to his camp while he'd been away with the old Indian.

The old man.....Walton glanced out through the crack in the half open door. The craggy faced old South American was there, blanket around his shoulders, half hunched against the tree. Crouched, just as if the body reflected the question in his mind. *'Where was the rain?'* The whole balance of the life he'd known had changed.

Once he could rely on the four o'clock monsoon, but now, if it came at all, it could come at any time. A rhythm of his life had gone. The wind, almost unknown when the jungle was thick, swirled away the smoke of his pipe.

His language had no word for dust, yet, these days, he could be covered in a split second. The blanket dropped from the old man's leg to reveal a sore. It had been there days now. Walton had tried to give him medicinal ointment, but the old man refused.

He'd wanted his own medicine -- some bark boiled into a paste. One of the young men had gone to get it, only to come back two days later empty handed. The tree he needed went the way of the dinosaur in the last bout of forest clearing.

Four thousand acres of trees burned in a single day -- and deliberately. It was one of thousands of schemes now pockmarking the jungle. Soon the cattle would be moved in -- but not for the Indians. The cows were the property of someone supplying hamburger chains. It was cheaper to grow the meat in the jungle than on the prairies.

Walton had been a witness to that last burning. He remembered the noise, the death cries of splitting trunks -- like useless limbs. Images then flooded his mind of Vietnam newsreels, charred bodies, napalm victims. But who would cry for the trees?

The animals too, enemies down the centuries, linked now by self-preservation, running blindly - satanic silhouettes against the flames.

And the flowers, some that Walton had never seen before. All this for cheaper hamburgers?

The sound of his helicopter brought Walton to his feet. As it took off, the old Indian, blanket clutched tight round him now, vanished in the dust. Minutes later Walton was looking down on the jungle, heading for Manaus. It was like gliding over a lawn that had been treated with creosote. When, if ever, would those brown patches be home to any life?

Two hours later he was on the Jumbo Jet heading North, row three as usual. Not for the first time he thought just how crazy this flying business was -- and how he hated it. Here he had no control. He felt dominated and, because of that, inept.

He'd always avoided being dependent on anything, drink, tobacco, even women for that matter. He enjoyed them all, but only when he had control. Here, he just felt helpless. He shared that with the old Indian ..... neither of them in charge of their destiny.

He re-checked his report. *'blind, unthinking, criminally irresponsible'* -- all the angry words stabbed their way through.
Maybe it wasn't the language of science, but what started as a detached, almost academic mission had accelerated rapidly into a blistering condemnation of human greed.

Once again it was the Indians who were suffering. Having swept them away from North America in one century the white man was doing it again in South America in the next. The UN agencies, like his employer, the FAO were just as impotent.

OK, so it looked fine on paper to have UN bodies and campaigns like "Man and the Biosphere" but what was the point when they got nowhere?

Half smiling to himself he thought of the rainforest vine in Mexico, now one of the prime source materials for the contraceptive pill -- what if the destroyers had got to that one before the scientists?

Unwittingly the thought linked him to Grace -- his lover and Foster's secretary. Tonight, he would be with her and he could hardly wait. But, for now, he slept.

***

He only woke as the plane dipped, starting its descent over Washington. Beneath them were the well-watered, well fertilised gardens and golf courses. Again, from the back of his mind came a statistic, something about Americans using more fertilisers on such places than was used for growing the entire food crop of the third world. It wouldn't have surprised him. Just knock out the commercials hinting that a brown lawn is a social disease and you're into a multi-million-dollar industry.

The plane landed -- and he found Grace on the airside of customs. Efficient, the super secretary greeting a returning VIP.

Walton lunged, to kiss her. "Dick" she muttered "not here ..."

"What's all this celebrity stuff Grace?"

She giggled "I'll explain in a minute."

Within ten minutes they were in her car, speeding away from the airport. "You see" she said "I had a word with the President and explained that you were extremely anxious to take a lady to bed and you couldn't wait to mess around with customs. OK?"

"Your bed or mine" laughed Walton.

"Your flat, not your bed and you'll have about ten minutes. Then you're going straight to the White House to see Joe Simon, the Vice President."

"OK" said Walton, suspecting a wind up. Another glance told him she was serious.

She explained "He's got a job for you. It's important and you start now, it's got nothing to do with South America or the FAO. From now you work for Joe and the White House."

Walton was silent. It was all too much.

They parked at his flat and went upstairs. Now they kissed, desperately thrusting themselves into each other. Reluctantly, panting, she pushed him away. “Easy boy -- we’ve got time. Right now, you’ve got work to do. Your car and driver arrive in five minutes and you need to change. You actually smell of smoke.”

“My what arrives?” Walton was having trouble keeping up with events.

“Joe Simon sent a car for you -- you’re obviously important now.”

***

Simon’s office was a mess. Papers everywhere. Posters fading from his Oregon days jumbled the walls. He talked fast, punching home points with both hands, walking around as if he stayed too long in one place the carpet would explode under him. Walton just sat and listened.

In three minutes flat, Simon outlined the problem with the returned space lab, its cargo of PWN and his worries about after effects. He fired a question at Walton “How does PWN kill trees?”

Walton explained “It’s pretty quick. It’s a parasite of beetles that bore into trees. Once the beetles are in then the PWN gets into the water channels. They’re like our veins, they carry food and water round the tree. The tree’s defense is to make dense blobs of a watertight gum and block the channels -- to stop the PWN travelling round the tree. But the tree’s reflexes are too slow. The PWN gets around the channels quicker than the tree produces the gum and soon the breeding of the worm itself starts to clog the channels as well. So, the combination of the blobs and the breeding of the worms stops any water or food going around the tree. In effect, the tree has helped to strangle and starve itself at the same time as the worms are attacking it. You’ve heard of Dutch elm disease?”

“Yes, it’s hit millions of trees all over the world. It’s the same?”

Walton went on “With Dutch elm disease it’s a fungus and in this case it’s a worm, but the effect is identical… Death.”

“But if it’s just pines...”

“No. It could also affect tropical trees. If it got into rainforests we’d be in trouble.”

“And it’s come down in Indonesia” said Simon half to himself.

“I’d need to go there to be certain” said Walton. “Is that OK?”

"Yes" said Simon "and don't worry about the cost or where you want to go -- this is kinda important. Remember I'm the ecology freak in this place."

"Yes, I know -- I thought that's what you wanted to talk to me about -- South America." said Walton.

"I do" said Simon "Tell me, what are you doing this evening -- where are you eating?"

"Well..." Walton hesitated, thoughts of Grace and bed flashing through his mind. Simon read his thoughts. "Don't tell me -- you've got a girlfriend you haven't seen for months. Right?"

Walton nodded sheepishly. "Sorry," he said, "but you know how women are."

"Bring her as well" said Simon. "She can talk to Ruth and you can tell me about South America. OK? That's fixed. A car will pick you up at seven, see you then. And hey," he grinned widely "I'll get you home early."

On his way out Walton borrowed a phone, called Grace at his flat and told her the news.

"And I'd dressed up for you too." she complained, half-jokingly.

"What are you wearing?"

"Your bath towel. Now I'll have to go home and change."

"Do you have to?" joked Walton "It's your second-best outfit."

They arranged to meet later. Walton was driven back to his flat.

***

She arrived five minutes ahead of the chauffeur's return. Walton gulped. At first glance the dress was simple. But it hadn't come from America, it hinted more at France or Italy. But still there were stockings and silk, the cool, classy look that so turned him on. Walton reached for her, cupping her breasts in his hand, then pulling her to him.

"Couldn't we say we got held up?"

She tore herself away. "Down boy. Later, we'll have all night."

"All night." He beat both hands on the back of his neck in not so mock despair. "Damn the White House."

***

In the FAO building Foster was fuming. His hand shook as he put down the phone. Not only were they actually taking Walton away from him, not only was he to have nothing to do with his work -- about which they were extremely vague -- but now to add insult to injury they were taking his secretary Grace to work with Walton as well.

If he'd been the resigning kind this would surely have been something to quit over. Foster, being Foster, had not betrayed his dismay, his fury, to his master. At least he sincerely hoped he had not. God Almighty, he hoped not.

He buzzed for Grace, to tell her how he'd decided she should help Walton for the time being. She was still his, of course, but in the circumstances...

Where was she? She could at least start the search for a replacement. The girl who answered was not Grace. She'd already gone. A heavy date, she'd said.

He wouldn't forgive them for this. He would, he vowed, get his own back. Meanwhile, what on earth was Walton up to? It had to be something big and, if it was, then there were people who should know. Like the ex-President's special friend, the one who ran the timber multi-national, the one who'd talked about that consultancy position Foster might expect to have in his retirement....

***

The Simon's house was comfortable rather than chic, possessions for use rather than ornament.

They ate simply, drank little and talked a lot. Dick explained to the two boys Peter and Joseph about the rape of the rainforest, their father listening, nodding agreement.

Peter, the older boy, knew much of the story from his father. "It doesn't stop there does it?" he asked Walton. "Doesn't it change the climate too?"

Walton agreed. Joe's wife Ruth wanted to know more. What was this greenhouse effect? Walton explained how he thought it was already with us. The weather was proving it all the time.

For the boys and Ruth, he explained how trees absorbed carbon dioxide caused by the burning of fossil fuel. Now, there were fewer trees to soak up the gas and the level in the atmosphere had increased dramatically. That trapped more of the earth's heat and radiated it back. Unless something was done soon the earth would heat up to the point where, said some scientists, the polar ice caps would melt, raising the level of the oceans and flooding coastal cities.

Joe also wanted to know about South America. How bad was it?

"Almost too bad for words" said Walton. "but, it isn't the only place. You've seen the reports. There are millions of trees vanishing all over the world. Millions. It works out at around fifty acres of trees every minute, every day of every week, every year. Everybody agrees it'll mean whole new areas of desert because there won't be roots to hold water and the rains will just wash away the soil. Then you can't plant food any more. It's a dozen, maybe more Ethiopias, all rolled into one. So, you'll still be getting more people in the world every day, but less land to feed them from."

Simon almost groaned. "How do you tell that to people?"

Ruth chimed in, "Floods already! Does nobody read the bible anymore?"

Simon left to take a phone call, returned and beckoned Walton into the garden.

"You look worried" said Walton.

Simon explained "That was the President on the phone. There was something else on the space lab. Acid. Tons of it. Highly concentrated. We've just found out."

"But it wasn't there to start with was it? How did it get there?"

Joe emphasized "We've only got the outline of the story so far. The details are in papers marked for the President only -- because Daley was involved to start with -- but what we do know is that it was transferred to the spacelab by one of the shuttles. It was a military one that was kept quiet at the time. Our guess at the moment -- and it's only a theory so far -- is that Daley planned to put something else on there later, but first he wanted a defense system. We think the acid was a defense against hostile astronauts."

"Shit" said Walton "what a crazy thing to do. The guy must have been a madman."

Walton's brain was racing now. "Did it come back with the spacelab?"

Simon shook his head. "We don't think so. We're not sure where it is. For all we know it could still be floating around in space. We'll know more detail when we get the papers. Meanwhile there's to be no talk of this. We keep our lips sealed, understood?"

"Absolutely" said Walton.

Simon fired another question at him. "You and Grace? How close are you two? Can you trust her?"

"What the hell's that got to do with it?" Walton was nettled.

Simon smiled slightly. "Nothing, but you'll need a secretary, someone you can trust. Would she do?"

"So, if there's any pillow talk you'd rather it be between the two of us. Right?"

Simon smiled apologetically. "Just a precaution. I've already told Foster she's the best qualified person for the job."

Walton grinned "I hope he never asks what the qualifications are."

They laughed then, the tension eased, though the problem remained. As they walked back into the house they finalised the arrangements. Tomorrow Walton and Grace would settle in at the White House; Grace would then stay there while Walton went on to Indonesia. The White House would issue a press release saying he was carrying out a wide-ranging ecology audit for the new administration.

They said their goodbyes and the chauffeur drove Walton and Grace to his flat. As the car took them back to Walton's place, through late-empty streets, Grace squeezed Walton's hand. "That, my boy, was a very impressive performance indeed."

"Bullshit" said Walton, though he too felt he'd done well.

In the elevator to the flat Walton slumped against the wall. He was dog-tired. He realised too that Grace, fresh and cool, was definitely turned on by this power thing. Was it some sort of Empress syndrome? Whatever it was he really didn't feel right now like doing his emperor bit.

He slumped on to the Chesterfield in the flat. Grace, still trilling, went to the bathroom. She emerged, clad only in a black negligee. Walton felt less tired now. She slid down silkily next to him; they kissed, and hardly able to help himself, he slid his hand to her partially tanned breast. The nipple came erect between his fingers.

Gently she straightened up, pulled him towards her and led him to the, bed.

"Still tired?" she smiled. She slipped off the negligee and posed for him. Walton was speechless. It was a sight to resurrect a dead eunuch. Hurrying now, he threw off his clothes and wordlessly drew her to the bed. Gently he stroked her breast and caressed her all over. Again, she moaned softly and soon her hands were behind his shoulders, urging him into her. She thrust against him, a gasp of ecstasy and soon her mouth was twisting in passion. She seemed to explode underneath him, a helpless shuddering and then he too followed suit.

Minutes passed before he could speak. Gently she kissed the side of his mouth. “I’m glad you don’t treat me like a lady all the time.”

Quickly she moved away, smiling almost maternally, pulling the covers over him. “I guess you’re tired. You’ve had quite a day.”

Walton felt like a log. He kissed her, then turned over.

“Remind me in the morning” he said “to tell you about my new secretary. She’s a very sexy lady indeed and I fancy there’s a lot of guys who could go off the rails with her.” Grace looked puzzled, said nothing and turned over.

Over breakfast, she reminded him. Walton smiled. “Well as I said she’s a very sexy lady. In fact, I’ve slept with her once or twice and I tell you she’s just incredible. Best thing ever.”

Grace could stand no more. “Listen...” she seethed. “Anyway, who is she?”

Walton burst out laughing. “She’s called Grace. Worked for some real pain in the ass guy called Foster. Of course, he’d have liked to...” He ducked as she laughed and threw the toast at him.

“You bastard. But is it really true?”

“It is.” But he didn’t tell her why.

***

The chauffeur collected them, dropped them at the White House and they moved into their offices. Within two hours it was all done. While Grace fixed his travel to, and reception in Indonesia Walton went to Simon’s office for a briefing.

By now Simon knew more about the acid. As they suspected it was a defense against astronauts checking the satellites of other nations. Simon explained it was a mechanism triggered by the heat of an astronaut in cold space. American visitors could immobilise it before going on board. But if someone didn’t have the key the acid would be released and eat through their space suits. They’d be dead in a minute. The Russians, or whoever, could never complain since they should not have been there in the first place.

Simon commented “The President’s talked to Daley and told him that one word out of place and he’ll be publicly exposed.”

*But what was he trying to protect? PWN didn’t rate did it?*

"We still think" said Simon "that he was going to secretively put something on there. It could only have been something to do with Star Wars. A laser maybe?"

"So, what do I do?" asked Walton.

"The first thing" said Simon "is to get you out of the way. Washington has too many reporters asking too many questions. We've got a press conference this morning, but that's just routine -- or should be. They've mostly forgotten about the spacelab by now. But there are some specialists who've been asking about a sudden change it had in orbit on the way down. Our guess is that it had something to do with the acid and maybe the PWN being dumped or released on the way down. Maybe the heat switch came on during re-entry and let the acid out. We're checking on it. We've got the spheres back and there's nothing in them."

Walton broke in. "Listen, I've been thinking about this. It's unlikely that the PWN would, if it's still alive, get into the trees that fast if the trees have their normal cover on. Now the pines in Oregon will still have that cover. What worries me is Indonesia. If the acid were to come down there as well then the leaves would be stripped, the PWN would have a perfect way in and God knows what would happen after that."

"That's why you've got to get there fast" said Simon. "Forget Oregon for the time being -- just get to Jakarta. We've fixed a cover -- you're just a White House advisor doing a routine tour for the new administration... increasing concern from the new boys here who care more than the others did, right?"

"I'll learn the script on the way" said Walton, unable to disguise the cynicism in his voice. Simon heard the sarcasm but said nothing.

***

He glanced at his watch, wished Walton luck and started to prepare for the press conference.

At radio station WWN the conference was top of the noon schedule. Bob Powell, though now detached from routine diary duties, had little interest in it. Ask a question and every other reporter heard it and could follow it up. What he needed was an individual story, an exclusive he could keep for next week's show. The interview with Simon had been fine and, as he'd hoped, Voice of America had bought it. He could do with another one. The better his status, the more he'd attract leaks from other politicians.

But this morning's conference was useless for that. What still buzzed around at the back of Powell's mind was Simon's remark about a can of worms. It wasn't a Simon sort of phrase. The President had called him about it and it sounded urgent. There had to be something more in it. And why appoint this guy Walton to do an ecology audit? A cover up? He could always check it out -- who knows what it might yield?

He decided to ignore the conference and went instead to the office bar. He ordered a bourbon and sat next to one of the station's new boys, Bert Casey. Mostly nowadays he was a news man, but his speciality was science. He asked him about the satellite story. Was it anything important or just another piece of space junk?

Casey smiled, flattered at the attention from a senior man. "Funny you should ask that. I was beginning to wonder."

"Why?" asked Powell.

"It's probably nothing" said Casey "just a couple of things that didn't quite tie up. I was picking the brains of one of the science buffs on a magazine the other day and they'd got some theory that the thing had suddenly changed orbit. It was almost as if it had lost part of its payload. They couldn't get to the bottom of it because they didn't get an answer out of the White House before their deadline. Last I heard they were still working on it, despite the problems."

"What problems?"

"They were just intrigued because when it went up all the enquiries were handled by NASA, as usual. But when they asked the other day they got referred to the Defense Department, not NASA. The guy they had on it isn't the world's best reporter, in fact he's a professor doing some moonlighting, and he just took it for granted they didn't know what they were talking about. Let's face it, he knew the Japanese were involved at the start."

"What did they have to do with it?" asked Powell, increasingly interested.

Casey shrugged his shoulders "I don't know. I mean it's all over now isn't it? The thing's crashed, nobody got hurt. End of story."

Powell gulped. To collect his thoughts, he went for two more bourbons and thought fast. An orbit that had changed, a piece of space junk, the Defense department and the Japanese. And, Joe Simon talking to the President -- about a can of worms. How was he supposed to make sense out of that? Was there even any sense to it?

He went back to Casey, changed the subject for ten minutes and made to leave. Just casually he asked Casey to see if there were any original clippings left at the science magazine on the spacelab's original launch.

"Why would you want those?" asked Casey.

"No reason really" lied Powell. "But if I was hard up I could always do an endpiece for the show about what would have happened if it had hit the White House. Just for fun, nothing more."

It wasn't until he left the bar that he uncrossed his fingers.

## CHAPTER THREE

The hot damp had Walton drenched in seconds. The airport tarmac shimmered a mirage and, for a second, he wondered if he wasn't back in South America. But this was Jakarta, capital of Indonesia. The army was everywhere. He guessed that sometime in the near future he'd end up dealing with them.

He was distracted by a soft voice behind him. "Mr. Walton?" He turned and involuntarily gulped. She was beautiful, like something out of a commercial for Singapore Airlines. But, with much more class. Dark hair, soft-eyed, slim too -- yet the instant impression was of a secret sadness.

This was a chauffeur? He couldn't believe it. Her English was perfect, though it had touches of formality that threw him.

"I am Lestari. My father thought it would be less troublesome if I met you."

"You mean your father's the Vice President Mr..." he stumbled apologetically over the metre-long surname.

She laughed, pronouncing it for him slowly. God, thought Walton, they can't possibly be all like this. He could hardly bear to look at her, yet whichever way he turned he felt he was staring at her. Down boy, he reminded himself. You're on business, and out here you probably get castrated for even thinking that way about the Vice President's daughter.

She slipped into her convertible and, as they drove from the airport, began to give him a guided tour. But Walton was making private notes: the shanty towns, people in aimless poses, youths lounging, old folk and children with all the pot belly hallmarks of poverty and malnutrition. And there were so many.

She answered his unspoken question. "Yes, we have too many people, too little work and food. It is very sad, but we are trying. My father tries."

For Walton it was an action replay of scenes from South America. Politicians there had tried -- they'd driven damned great highways through the jungle, roads that led to nowhere but starvation.

The hotel was the usual, enormous modern block. He asked if she'd join him for coffee. She smiled slightly, shook her head gently and politely refused. "I am the Vice President's daughter and we still have a large Muslim community. Perhaps another time." She continued

"You will want to bathe and rest. My father has extended an invitation to you to take some food with us this evening if you are not too weary. Would that please you?"

Walton nodded, mumbled thank you, feeling like an embarrassed schoolboy with a crush. She left, telling him he'd be collected at six. Only then did he realise he was standing on the pavement with a foolish grin on his face. What a lady!

He checked in, stripped off and showered. Adjusting his watch, he realised he just had time to call the office and Grace. She'd like that. Grace took the call. Walton, disguising his voice barked "You just get me that son of a bitch Walton. My name's Foster and he's just hightailed it off with the sexiest damned secretary I ever did have. She had a figure like…"

"Hi Dick -- you sound great."

I'm OK, but it was a long flight and this room's feeling kinda empty already. Do you know I had to take a shower on my own?" She giggled, knowing what he meant.

Listen, sir" she said emphasising the 'sir' "I have news for you. The press are after you. A radio reporter named Powell called the flat. I only answered in case it was you."

Walton felt a slight chill. "What did he want?"

"He just said he needed to talk to you about your new job. He's got a show called 'Folks on the Hill' and he thought he'd do a piece about a newcomer. I just said you were out of town for a few days, but I didn't tell him where. He wanted to know what you did, and I said your speciality was deforestation and its effects with particular reference to South America. I guess he then presumed that's where you were."

"Well done darling -- I always knew you were a good secretary." They said goodbye, inhibited by the thought that the call might be bugged -- at both ends.

***

At six precisely, to Walton's surprise, a chauffeur was waiting. He was in Army uniform which might have explained the punctuality. They cruised out through the quietened city and beyond past the teeming shanty towns. Soon the road opened up. As darkness collapsed on them, with the usual tropical suddenness, he saw only silhouettes of the luxurious vegetation.

The Vice President's house, armed guard at the gate, was large, low and cool. It was sparsely furnished and had an oddly colonial feel. A touch of Noel Coward had filtered through somewhere. Yet, it still felt like a home.

Lestari, now in traditional style, had a simple white lace blouse which gave hints of a smooth, tanned body beneath. Even as they made their re-introductory remarks there was an uninhibited yell and a little girl, around three years old, hurtled into the room.

Lestari smiled apologetically. "This" she said "is Nani. Noisy Nani. She's very excited because I told her you'd come all the way from America. She's used to diplomats and the like, but they usually come in groups and, because you've come on your own, she's somehow got the firm idea that you are different -- that you've come especially to see her."

"May I?" Walton leaned down, picked up Nani, tossed her into the air and caught her again. She squealed with delight.

"Nani" he said very gravely "I have come from America to see you because your Mummy tells me you're a very good girl. And sometimes not so good. Is that so?"

"I'm a good girl" said Nani solemnly.

"Hey," said Walton, conspiratorially "I bet you've got one very special doll, haven't you? You see I know about these things. Now her name is... let me think, let me just try and remember..." Walton looked over Nani's shoulder at Lestari for help.

Lestari mouthed *'Kim'* and Walton scratched his head. "I believe it's Kim."

"That's right" Nani looked at him wide-eyed.

Walton went on "Now what's so special about Kim. Let me try to remember."

Nani came right out with it. "My Daddy gave her to me."

"Of course. I remember now. Now what is it your Daddy does?"

Lestari's face dropped. Nani looked at him in scorn. "Don't you know my Daddy's dead?"

Walton felt sick. And bleak. He turned to Lestari. "I'm so sorry. I didn't know." Lestari shook her head, moved forward, quickly bent down and picked up her daughter. "Oh Nani, he's a silly man isn't he? That's something he just forgot for the moment. Go and fetch Kim. Show her to Mr. Walton."

Nani ran off. Walton turned to Lestari "I don't know what to say. I'm so sorry."

Lestari, recovered now, put a hand gently on his arm. "Do not worry. You could not have known. He was in the Army. Are you married Mr. Walton? Nani really took to you and it looked as though you liked children. Have you any?

"No and no, but I do love kids. And Nani is enchanting."

She smiled.

"Did you re-marry? Or do you plan to?"

"No, I didn't" Lestari said "and I think perhaps I won't. Though I think Nani would like it."

Nani rushed back into the room, holding up Kim for Walton's inspection. "I think" said Walton giving the matter his most solemn consideration "she's almost as pretty as you are. Almost, but not quite." In a sudden flurry Nani jumped on him flung her arms round his neck, then kissed him on his cheek. Then she dashed back to Lestari and half hid behind her.

"She is very impulsive" apologised Lestari.

"Don't worry" said Walton "it's not every day that happens to a fella." His voice had a tinge of gruffness. It was like he had a golf ball in his throat.

Lestari took Nani off towards the door to begin the bedtime ritual. She paused "My father will be here shortly. Would you care for a drink? I should have asked before."

Walton could have murdered a sizeable slug of bourbon. "A long lime juice would be wonderful, thank you."

Within a minute a servant had appeared with it on a silver tray.

Walton settled back in a rattan chair, toying with his glass. The last few minutes had got under his emotional belt, left him with feelings he couldn't quite understand.

Sitting there, dreaming, he was brought back to earth by the entrance of an old man.

Lestari's father, slightly stooped, had a face fringed with a trim of white beard. There were lines of care and of age. And, thought Walton, wisdom was there too. And something else - dignity.

"Mr. Walton" the old man said "I must apologise. I had meant to greet you. Forgive me. But I gather the younger generation deputised for me rather well."

"I was made very welcome."

"Now" twinkled the old man, eyeing the lime juice "are you sure you would not prefer some Scotch whisky."

"That would be wonderful sir." Moments later a servant handed it to him. They sat and went over briefly the purpose of his visit.

The care lines in the old man's face creased deeper as he talked of the sphere that had landed. "Your people were quick to remove it, but I'm afraid some damage seems to have been done. The army has closed the area and we have arranged for the press not to mention it, but I'm afraid there is a problem.

Walton's heart sank. The old man went on "There is damage over a wide area which cannot be explained simply by the arrival of the sphere. Many trees have become defoliated and I have to ask you if there is something we should know. Is there Mr. Walton?"

His eyes were piercing now. Walton faltered "Mr. Vice President, my job is simply to investigate and to advise and..." he faltered again, "to report back to my Vice President." The old man stood up, less stooped now.

He said, "I am reminded by my President that there are those in America who think we have been a little harsh in the treatment of some American firms and that perhaps on this occasion our help would redress the balance."

Walton knew what he meant. Several US based multi-national timber firms had moved in, invested millions of dollars then found the government insisting on a larger and increasing stake in the business and, while demanding a 35-year harvesting cycle, would only grant them 20-year leases.

They weren't happy about it; and chances were in a few years' time they'd simply pull out and leave the Indonesians to it.

The Vice President was still standing. He went on "As I said we wish to co-operate, but we can only do that in partnership with you. That implies honesty on both sides does it not?"

Walton was embarrassed. Was he a fall guy? Nobody had prepared him for this. He gulped. "Mr. Vice President you must understand that I am merely a scientist, not a diplomat nor even a politician. My brief is very clear."

He didn't get away with it. The old man was sterner now. "Mr. Walton. We are not fools. We too have scientists. They tell me the damage at the landing site could not be that of the impact alone. If we have knowledge we can be of help. Without knowledge..."

Now Walton rose to his feet. "Mr Vice President, you must understand I have a loyalty to my government. However, I do agree that honesty is necessary. Will you take my word I shall do all I can to ensure I am authorised to tell you everything I can?"

"You'll call Mr. Simon tomorrow?"

"Yes sir, I will." Almost sensing a cue Lestari came back into the room. The tension eased.

She smiled at Walton. "Nani included you in her prayers. You obviously made a good impression."

Walton smiled a thank you. "She's lovely. Thank you for introducing us."

They moved into dinner and, when polite conversation faded, Lestari asked about his work in South America. He couldn't help it, he was conscious of it, but as usual as he got on his hobby horse his anger rose. With some passion he talked of how the South American Indians were being exploited by big business.

The old man was all attention, but quiet. Walton realised he'd maybe gone too far and apologised.

But now, the Vice President had a new tone in his voice. "It is good to hear not just what you say, but to know how you feel in your heart. It is in accord with us, though it makes me sad that some of my colleagues see the current wealth in our timber as a chance for personal gain."

Walton looked puzzled. Lestari explained "What he means is that corruption has become part of our lives here. We have tried for years, but it is difficult to change a way of life."

"What sort of examples are there?" asked Walton.

The old man became more animated now. "There is one that is close to your heart. Because of the way the commercial system works, the companies -- and the ones affected are rather more Japanese than American -- are obviously anxious to get as much timber from the forest as they can. They pay their staff well by the standards here and on some days -- perhaps most days -- you will find that if one of our foresters is at the scene of some cutting he will notice that too many trees are being taken. Now the bulldozer operator or the tree feller will be earning twice as much as that forester of ours. We can't afford to pay him much and we don't have many of them -- not enough by far. So, if he is offered a little inducement -- we call it a 'pungli' -- he'll go away for a few days and study another part of the forest. Who is he to say 'no'? Does he care more about the forest than his family? It is sad, but what can we do?"

Walton had a short answer "Surely you can explain that if it's not looked after it won't be there in a few years' time."

The old man smiled patiently "But it is not just the companies that take the trees. For many of our people the jungle is their home. They need the trees, they clear them to grow food, they need the wood for cooking. Can we say to them 'No -- these belong to the Americans and Japanese'?"

The discussion went on long into the evening, Walton asking about the government's policy, the old man spelling out for him the facts of life.

For the old man it was a dilemma. His country was the world's biggest producer of the highly valued tropical hardwoods. They were offered millions by the Japanese on whom they depended for a lot of trade and by the Americans who they needed for defense. The Indonesians had something to sell but not the knowledge or the money to do it themselves. To get the money they had to invite in the multi nationals whose main concern was the stockholders back home. Within ten years one whole forest had vanished.

It was early the next morning before he left for his hotel. After his promise to call Simon the old man had promised him army transportation the next day to the spot where the spacelab had crashed. He dreaded to think what he'd find.

***

The hotel room was like all hotel rooms. He thought about phoning Grace, but it was probably too late now. And what could he tell her? She probably knew more about the forest there than he did -- knowing her she'd have been reading the reports. And what else could he tell her?

He could hear himself. "And this woman, Lestari, she's so beautiful it knocks your eyes out. And this lovely little kid Nani....."

***

Bob Powell had put in a hard night. His expenses gave the impression he spent large chunks of his life -- and the office cash -- in the bars around the Capitol. They were right. In a government town like Washington the bars were the source of all good information.

Like all other towns and all other bars, reporters got their stories that way. Back stabbing politicians dished out the dirt and ambitious officials leaked their pet projects -- or other people's. Nothing was ever attributed to anybody, the first reporter even to hint at a source found himself facing a line of backs.

Tonight, trade was slow. Two bars and two bourbons into the evening Powell had only heard about an under-secretary screwing a senator's wife and a congressman facing an IRS grilling over "consultancy" fees.

It was time for a night-cap -- maybe he'd have one for the road in the bar used by most of the UN crowd. It was late now, and the shop talk was exhausted. By now the talk was all about girls and who'd laid who. The bar crowd was drinking a toast, a wake, for the departure of a secretary with the best legs in the building -- some FAO girl.

"Didn't some guy called Walton work there? Did anyone know what he did?" Powell made the enquiry casual. Walton had gone now, said one of the crowd, hightailed it off a week ago to the White House, same place as the girl. Happens all the time, nothing in it at all, said the crowd.

Powell played it dumb. "But they said it was *'a wide-ranging ecology audit'* so why hire a forestry guy? Lean doesn't give a damn about trees"

"No, but Joe Simon does -- maybe he's working for him."

Powell pretended to lose interest. "Now this girl -- she a new England type, classy sort of voice?"

"That's it pal, real class. Name of Grace. I heard they were an item, but you get that about every guy and his secretary don't you?"

Powell nodded, bought the man a drink, chatted about baseball, then left.

Grace, if that was her name, had to be the girl who'd answered the phone at Walton's flat. She sounded classy, his radio ear put the accent in the Boston area, and she knew about Walton's work. Now they were both in the White House and in a hurry and, possibly, both working for Joe Simon, the ecology nut. Simon had talked about trees -- and that *'can of worms'*. And it had sounded serious. He jotted down all the details he could remember.

He wondered if Casey had the clippings on the spacelab's launch. Powell couldn't work out why he thought the spacelab and the trees were connected. But his gut instinct told him it was so -- and he trusted his hunches.

Casey, next day, had the clippings. Powell took them off-handedly, thanking him playing down his eagerness to look at them.

When he did he was confused. What the hell was a pine wilt nematode? He read on, discovering in a paragraph from a tabloid the description he was after -- a Japanese worm that killed trees.

He almost laughed out loud in sheer delight. A can of worms...

So, if Walton was working for Joe Simon, who he knew had seen the President about a can of worms, his work had to be connected with the spacelab. A few pieces of the puzzle were in place, but now he had enough to go on.

He'd have to bluff, but he'd done that before. It was worth a try.

He moved back to his desk, called the White House and asked for Mr. Walton's office. The New England voice answered, he was sure it was the same one. "Is that Grace?"

She sounded puzzled. "Who's this?"

"It's Bob Powell, from WWN, the radio station. I called you the other night remember?"

"Oh yes, Mr. Powell, can I help you?"

"I don't know. I wanted to speak to Mr. Walton. Is he back from Indonesia yet?" It was sheer bluff.

"Who told you he was there?" asked Grace. Powell lied.

He gulped "I don't know. I guess it was our people in Indonesia. Or maybe it was South America. I just got a note saying he was here -- but is he back yet?"

Grace hedged. "What was it you wanted to talk to him about -- can I help you?"

Powell paused, she sounded delicious. He hated giving her a hard time. But he also had to find out. "Listen" he said "When he calls you could you tell him I'd like to ask him about PWN -- you know pine wilt nematodes? Oh, and the spacelab. It's what Joe Simon called the can of worms."

Grace sounded worried now. "I think you should talk to the press office shouldn't you? I mean, that's the procedure."

Powell relented. He had enough now. There'd not been one iota of denial in her voice, Walton was obviously abroad and the reference to the press office always meant a department was fending off an enquiry.

His voice softened. "I guess you're right. I'm sorry to involve you in all this, but you know how it is...."

"Sure, you've got a job to do. I know that."

Powell laughed. "Hey, listen, I don't work all the time. I mean if you had time for lunch or even a drink sometime. Joe Simon will vouch for me -- a clean, upstanding guy, no vices, treats ladies nicely, sound in wind and limb."

Grace laughed. "Mr. Powell, you are very persuasive, but I might just have a date you know."

Powell spoke without thinking. "Yeah, but if he's away..."

Grace hesitated. Just how much did he know? She'd been at Walton's flat when he phoned first time and he was obviously putting two and two together very fast. She could put him off the track by having a drink with him. What harm would there be? She might even find out just how much he knew. And Dick was away -- and Powell did sound fun.

"Mr. Powell -- two and two can make five you know. I might just take you up on that offer."

She'd done it now. Powell was astonished -- and slightly baffled.

"How about dinner. Pick you up at seven?"

Before she could think about it she said 'yes' and put down the phone. She picked it up again instantly, calling Joe Simon's secretary. She briefed her on Powell's questions, her suspicions that he might speculate about PWN and the spacelab. She didn't mention she was meeting him for dinner. But she wondered how Walton would react.

***

The alarm exploded at 2 a.m. For Walton it was an earthquake. In his dream Lestari, face down on the white beach, was about to turn to him. Already the top of her stark white bikini had been undone. Acutely, he was aware of his right thumb and forefinger. Any second now and they'd caress an erecting left nipple.

But the earthquake kept going. It was the bloody phone. Why was there a phone on the beach? Walton reached out -- the nipple was a handset. A voice was coming at him from somewhere.

"Mr. Walton? This is the White House?"

"Shit." And why did these stupid broads always put a question mark on the end of a sentence? Was Lestari really gone? He searched through his mind's eye. She was still floating out there, dimmer now, her brown body whirling, turning towards him.

"This is the White House? Is this Mr. Dick Walton? I hope you're having a good day?"

"Uh?" Walton believed it now. He grunted again, muttered "Shit, hold it" and fumbled for a light. It came on. He winced and threw the phone away to the pillow. It was 2 a.m. The handset was still talking. "The White House sir? Mr. Walton?" He grunted, more awake now. "Mr. Walton are you there? This is the White House Mr. Walton? Really sir, this is an intercontinental call? We do have a budget situation."

"Crap" growled Walton.

"Putting you through to Mr. Simon now."

Walton woke up at that. He made it short and simple. At that time in the morning he was capable of little else. He told Simon the bad news -- that the damage, defoliated trees and the rest was worse than expected purely from the impact.

And the Indonesians had guessed something was wrong. The question was how much could he tell them?

Simon too kept it short. "Tell them the least you have to. I'll have to leave that up to you -- you're the man on the spot. But whatever you tell them could leak and there'd be a fuss we don't need. Whatever's happened there could have happened in Oregon too. We've already got one reporter nosing around -- he's already spoken to Grace twice by the way."

Walton interrupted. "I knew it was once -- she told me. But surely she didn't tell him anything?"

Simon snorted "She didn't have to. He seemed to know, or at least had guessed most of it for himself. He certainly didn't believe the hand-out. I guess it's bound to come out, but we need time to get ourselves organised, to show people that if there is a problem then we're on the case and doing something about it." He paused, "By the way don't forget to stress that it was Daley who put the worms and the acid on board -- not us. OK?"

"Right" said Walton.

Walton had a question. "Did you find out any more about the acid?"

Simon said "John got on to Daley when he read the papers on it. Apparently there was a plan to put some military device on board, but they never went through with it. We're almost certain that the acid and the worms were released on the way down somewhere."

"So, they're floating around in space somewhere?"

"That's right" said Simon. "NASA reckons they came out just before landing so if they have come to earth it's probably near the site. We're checking on stratospheric currents and all the rest at the moment."

Walton whistled "Oh Jesus, that's a disaster. You realise just how bad this could be don't you?"

"I'm beginning to Dick, but I don't think it's fully sunk in with John yet. Are you OK otherwise?" Walton said he was, then asked about Oregon.

"That's your next stop" said Simon. "I'll explain to Grace, don't worry. She'll be busy anyway."

The call ended, Walton lay back on the bed. Christ, what a mess. Tomorrow could be worse. He was flying with Lestari to the landing site.

## CHAPTER FOUR

Walton met Lestari at the airport, the military section. Their plane, an aging Islander, cruised low and slow over the jungle.
Walton could hardly believe his eyes. As far as he could see, towards the shimmering horizon, there was an ugly brown stain replacing the normal dark green. It was like a raided graveyard. Just skeletons were left.

That was it, he thought, they're skeletons. Yet he knew they'd still be living -- for the moment anyway. His face paled. Blind anger, bitter regret at being part of the human race that could impose this fate on a fellow species of the planet. What right had they? What God-like quality did humans have that allowed them to do this without penalty? He thought again. Without penalty? There would be that alright, but not for this generation.

Lestari was pale and, very softly, weeping. "Oh Dick. Oh Dick. What have they done?"

They landed on the jungle strip at the Western end of the Kutai reserve, near where one of the spacelab spheres had crashed. But, as they flew over the eastern end, the much-vaunted nature reserve now being vandalised by loggers, they could see it was the same there. He took a gun from the plane and walked with Lestari into the jungle. She asked about the gun. He explained that with this amount of disruption the prized orang-utan population could be affected. They could be wounded and might be dangerous. She nodded and moved closer to him. Comfortingly – and glad of the excuse -- he put an arm round her shoulders.

Walton had never been in a jungle like it. Instead of the normal half-light through the canopy, the sun was now blazing down, baking the shallow soil. And it was so quiet. The birds who'd once weaved their private mysterious patterns under the canopy had moved away. But to where?

Here and there, already dried up water channels had started to show. With no canopy to stop it the monsoon had come beating through straight to the floor, collecting in hollows, overflowing and sweeping everything aside to find a lower level. It was the classic birthmark of a desert, new-born.

The leaves might grow again, but already the fragile eco-system, complex beyond human understanding, had probably suffered a mortal blow. His anger made him feel nauseous. All this was unique, irreplaceable.

For all he or anyone else knew they could have just wiped out the habitat of some small plant that could have held the cure to some major disease. They'd found quinine in the bark of a South American tree in just such an area. Now it was all in dreadful danger.

Ahead of him he heard a rustle. A pair of tortured eyes stared at him. An orangutan, half of its body fur gone. There were probably others. Quietly he turned and they walked back out to the airstrip. Lestari looked up at him, silently questioning.

He could hardly bear to look at her. "I feel like a firing squad looking at a line of babies." He flopped in the shade of the wing and made some notes. Stark, scientific, professional. This was only the effect, not the cause. It was now evident that as the sphere came to earth it had trailed behind it for miles a stream of acid and maybe some PWN too.

If the PWN came down here then, pines or not, the worms would almost certainly thrive. What the acid, the burning sun and the erosion had started, they would finish. And, he noted, there was no known cure.

***

Wearily he climbed back into the aircraft. Lestari was still silent. She knew he was suffering enough. They took off and this time Dick could hardly bear to look. With a few words they agreed he would have to see her father before he caught the night plane to Oregon. And, Dick reminded himself, he'd have to call Simon.

The old man read the story written on Dick's face. And his first question showed his bitterness. "Did it have to be us?"

"I'm sorry sir, we had no control. I think, sir, there's another question I'd ask. Did it have to be anybody?"

Walton gave him his report, told him now about the PWN and the acid. There was little point in hiding it.

It was like a carbon copy of the report the old man had received from all over his country. He handed Dick copies. Defoliation, the start of erosion; all there a dozen times over. There was nothing more to be said.

***

Walton re-joined Lestari who was playing with Nani. The little girl rushed up to him, they exchanged heart-stopping hugs, then suddenly Nani drew back and looked up at him. "Why aren't you happy today?"

Walton forced a smile and tickled her to make her laugh. "I'm OK now I've seen my favourite girl."

Lestari sent Nani, protesting, on an errand and walked to the car with Walton. She told him "It's not your fault you know -- you couldn't have done anything about it. Does this mean you won't be back?"

He shook his head "No -- I'll be back. There will be a lot to do in years to come."

She smiled "I'm glad", adding hurriedly "and Nani will be pleased too of course."

"Of course," said Walton and they each knew what they meant. He offered his hand, she took it and they held on. It was another physical contact and Walton at least was loath to break it.

Walton was suddenly aware of the chauffeur's curiosity and let go Lestari's hand instantly. She smiled and whispered, "You're learning our ways."

***

He left and went back to the hotel. He called Simon at home, thinking it was probably more secure that way. He told him the scale of the disaster and asked if Powell had used his story yet.

"No" said Joe "but I guess it's only a matter of time. Get some sleep on the plane, have a look at the situation in Oregon, then call me. We do have initial reports that it's much the same there, but I'd prefer you to have a look, then you can compare things. I'll call John with the news. Grace sends her love by the way -- I talked to her today. Don't forget to call her will you?"

"I won't" promised Walton and said goodbye.

He lifted the phone again to call Grace. While he waited for the operator to make the connection he poured himself a large one and returned to lay on his bed, his mind racing over what could be done.

The phone rang -- it was the operator. There was no reply from her number. He wondered where the hell she was.

***

It was an oddball restaurant -- plain pine tables, a suggestion of sawdust on the floor, red and white checkered tablecloths and the antiques hung discreetly round the walls just happened to be genuine. The waiters matched the antiques, but only in their discretion -- they'd seen too many Capitol Hill tête-à-têtes to be surprised at anything. And Powell had used the place before. Grace sensed it, but, after days without a call from Dick, just messages relayed offhandedly through Joe Simon's secretary, she felt used. OK, so work was important, but she was a lady and his lover, not just a casual girlfriend.

Powell was bubbly, charming, strong on the jokes and the manners. She was disarmed. The talk was of horses, the theatre, clothes and design. And, as she suspected, it changed with the coffee.

He was engagingly frank, apologising for hassling her, making it plain he now knew all about the PWN. She couldn't deny it -- he'd been clever enough not to ask her.

But, he knew that if he'd got it wrong she'd have told him, put him off, tried to confuse him by diverting him to another theory. He asked no questions, she had to give no answers, and, by the end of the meal, he'd confirmed for himself his theory was right. He charmed her too on another level, one that reminded her in a way of Dick. Like him Bob was dedicated. With Dick it was the environment, with Bob it was the truth. With this she could agree.

And with him she could gossip about any other UN or Washington department other than her own. With the brandy he became passionate about the public's right to know the truth. She was impressed. He meant it -- she admired and respected him for it. He was even a gentleman. At the door of her flat he just shook her hand, held it for only a fraction longer than necessary and said next time he'd like to be the cook. If she was free, that is.

"I could be. Who knows?" she smiled.

"Depend on anyone I know?" he teased.

"Oh yes" she said, "On me." They laughed, he left and by the time the door had closed he was driving away. He pulled around the corner, into the first parking lot, retrieved the notes he'd made during a spurious visit to the john and added some more. Her very lack of denial on some key questions had convinced him he could now use the story.

***

Next morning at work it was a slow news day. He had the answer to the news desk's problem, a summary of his story telling how there was PWN on the spacelab -- bugs in space he called it, how Joe Simon had hired a top scientist to fly to Indonesia to assess the damage.

It was just a trailer for his own show that night "The Folks on the Hill." In that he expanded it all, weaved in comments from ecology groups, hyped it into a major national political issue. The end question was critical. "What – if anything -- was the President doing about it?"

John Lean was not amused. Noon editions -- even before Powell's weekly show -- were all headlined "Bugs from space." It was as he'd expected. Next day's papers -- the Washington Post and those taken seriously abroad -- would be full of it. CNN was already running it hourly round the world, now getting into the background issues, asking all the awkward questions.

Lean felt uneasy --and angry. He'd already called Joe Simon. Simon replied curtly that he was fending off angry calls from Oregon based timber multi-nationals who were pointing out just how much they'd put into campaign funds.

It was getting ugly. Lean called Joe again, putting down the phone as Simon appeared at the door.

"Joe, how do we handle this?" Gone now was the forced air of calm, the laconic ease for which his PR team had made him famous.

Joe dived for the coffee. "I talked to Walton last night -- you know, the guy we sent to Indonesia. It couldn't be worse there."

Quickly, he gave him the headlines, the news of the acid and the PWN, the devastation now spreading through Indonesia.

Lean fumed. "What answers do I give Joe? What do we tell people?"

Joe paused. "John, it could be worse. Don't forget they only know part of it yet. All they have is some half-baked radio story about some worms that may or may not be alive. They haven't even heard about the acid. Maybe the worms aren't alive. If they are then maybe they won't last that long. Maybe the acid gets diluted after a while. Even if it doesn't there's still acid rain. That's been a problem for a long time.

We can afford to stall until I hear from Walton. He's on his way to Oregon now. We should know in a day or so."

Lean didn't look mollified. "By then I'll have the Russians on my back. Just don't forget, when I got here, I called Lentov and told him we meant it about being 'Clean with Lean'. You remember all that open government stuff we promised?"

Joe nodded. "I do. I agreed with it. Still do. I'm not even certain it wasn't my idea in the first place."

Lean burst into a laugh. "Joe. You're right. It was. And don't you ever forget you believed it at the time. What are you going to tell them now? You really going to say we put up a spacelab with bugs on it, then acid and the next step was lasers? You know what that would do?"

Joe paused. "OK, so we can still fool some of the people some of the time. We tell them Daley, and we stress it was him -- put this experiment on board -- everybody knew about it then and nobody objected then. Only now it's crashed, and we've got a top guy seeing if the bugs are even still alive. When we know, we'll tell people."

Lean winced. "If I was in opposition I'd have you for breakfast. But you're right it's the stall we need. Just do what you can with it."

Simon hurried out, still clutching his coffee. Stopping by to see Grace he asked, "Have you heard from that man of yours?"

"No, I haven't. I'm hoping he'll call me from Oregon."

Simon thought for a moment. "Look I need to talk to him. Can you see if you can get him for me? I don't know where he is exactly, but you know where the site is. Either the army or the national park service should be able to get you through. The chances are he'll be out on the ground, but they can call him on the radio. Just get him to call me."

Simon scuttled off. He could easily have had his secretary make the call, but Grace seemed worried that Walton had not been in touch. He had an idea Walton was more in lust than love, but it was just a theory.

***

In Oregon the track wound Walton up to the bare crag of the peak. He killed the engine, sat there and gazed around him. Below him the valleys, black shadowed in the early light, stretched three ways. In all of them a sullen grey mist swirled sadly, parting now and then to reveal a spiky graveyard. The slender pines, so proud and tall, were merely skeletons now.

Where once the winds had bounced the mists along their crests, the gusts dipped down now, chilling the trunks. Just here and there, in the lee of a hillock, a fully clothed pine pointed up the poignant contrast. He felt like a mourner, paying his last respects to an old friend, knowing now there was nothing more he could do. He sighed and grabbed his gun.

Outside the chill surrounded him as a small cloud of mist swept up over the crag. For a moment everything was blotted out. All was grey, all was cold and, as it enveloped him, the eerie howl of a wolf wailed a lament in the woods below.

He shivered and tightened his grip on the gun. The animals too were being hit. Creatures which, by now, should be deep in hibernation were still on the move. Their world too had changed. Gone was the cover that would gently filter the snow through the forest roof.

The mist drifted away to unveil a nightmare. It made him gulp. He felt bitter and sad all in one, but neither tears nor anger would clothe the trees again. He shook his head, bringing himself back to the practical scientist he was.

All the efficiencies of the laboratories, computer programmes in the experimental station back in Portland were fine, but he had to be out on the ground today. He needed to feel the tragedy in his bones. While science might identify the cause, he had some illogical feeling that only here would he get to the truth. It was almost as if he wanted the trees to divulge some secret remedy of their own.

Fancifully he wondered -- if they could talk -- what would they say? They'd have to ask what it was they had done to be treated this way. For thousands of years they'd been the servants of man and animal, freely giving of life and sustenance, warmth and cover, fuel and sanctuary. And man, in his own tiny way -- at least in the recent minutes of the world's hour -- had acknowledged that debt, trying to protect them from fire and disease. They'd had a deal, but mankind had welched on it.

The growl of the bear was close. Walton stiffened, alert now and nervous. Was this the place to be? The animals too would be angry with man -- and with just cause. Man had just stripped their homes of their roofs and emptied the larder. Now man as well was about to pay the price. He'd survive -- he always had. But he should atone for his sins. The thought brought Walton back to earth. He wasn't here to dream, to philosophise or to blame others; he was a scientist with a task. Taking a small axe from the truck he selected a small pine. Stupidly perhaps, he first touched its trunk, silently apologised, then sunk the axe into its bark. The V shaped wedge came clean away. The rings were clean, a sign of health.

Gently he touched it with his forefinger and brushed it along the path of the rings. It was sticky, not with normal resin, but with a heavier gum -- tylosis, the tree's last attempt to gum up its own waterworks and prevent the spread of the PWN. There was even one in there. He picked it out on his finger, hurled it angrily to the ground and screwed it into the soil.

Demented almost, he stamped and stamped on it. Anger overwhelmed him. Angry tears now stung his eyes. The ground was a blur. And still he stamped. "You bastards" he cried. But he wasn't quite sure who he meant. He felt sick and angry and helpless. And, for almost the first time in his life, it was as if he wanted to pray.

The trilling of the phone in the truck interrupted his thoughts. He lurched into it, grabbed the instrument and growled angrily. Was there no escape?

The phone went dead. Walton slumped into the seat, hunched forward, looking out through the windscreen. In the far distance a wraith formed itself almost into a question mark. The phone went again. He grunted into it, distractedly. It was Grace.

"Hey are you alright?"

"Yes". He was curt, still gazing at the ghost-like query in the sky.

"Of course, I am. What do you want?"

"Dick. This is Grace, honey. What's the matter?"

"Every bloody thing in the world is the matter. That's what. The whole damn place is sick and there's nothing we can do about it."

"Dick are you OK. Have you been drinking?"

He calmed down slightly, apologised. "I'm sorry honey, but I'm on a mountain top and it's the worst sight in the world. Even God just put up a question mark and I don't know the answer."

"He did what?" "Never mind. What's the matter? Is there a problem?"

"Dick. Joe Simon wants a progress report. Can you give him any news? He has to see the President again today. And the press office needs some information too."

"Honey, just tell Joe it's as bad as we feared. It's hopeless. It's finished. We don't stand a chance. And you can tell the press office if the reporters want to know just how stupid mankind could be, they could just drag themselves off their asses in those bars and get up here to look for themselves."

"They can't Dick -- the President's just made it a restricted area."

"Then maybe he'd like to tell that to the worms."

"Dick, when are you coming home?"

"When I've got some sort of answer. That's when. I'm not going to solve anything at meetings in Washington. I have to be here. Just tell Joe that. At least he'll understand."

"Dick, so do I. You don't have to take it out on me."

"Honey I'm not, but I'm working. And it's important and I have to carry on doing it. Can't you get that into your head?" He paused. "Honey I'm sorry, but you can't believe what it's like here. Nobody could. I'll call you when I can. By the way I called the other night and you weren't there."

Grace bit back. What arrogance! She blurted out "I went out to dinner with Bob Powell, the reporter. It was for your benefit, you know. But you might like to know it was fun as well. I got treated like a lady for once and I'd forgotten how good that could be."

"Honey, don't you know I love you?"

"How could I? When did you ever tell me?"

Walton apologised, but irritated now and it showed. They said goodbye and in seconds Walton was focused again on the outside world. That was more important at the moment. Grace was fine, but if it came to it there were other women and he had a job to do. How could people not realise how serious this could be? Didn't they think? He focused again on the outside world.

The question mark was breaking up, but the outline was still legible. "Yes God" mused Walton. "No wonder you've got questions."

Driving back down the track he switched on the radio to catch a news bulletin. The item was there about the area being declared a restricted zone -- as if man could do anymore damage, thought Walton.

The bulletin was followed by a public service announcement. "And don't forget folks if your radio uses a battery then you can get extra life from it by warming it when it seems like it's worn out. Batteries have paper coverings and if those worms eat all the trees we might not have paper much longer."

The DJ came back on the air. "And hey folks have you heard the very latest from the fashion scene? And it's for guys. Just think when the dollar bills run out I guess we're going to have so many coins we'll have to carry handbags. Now ain't that just fun? As it happens there's a firm that's already churning 'em out just for guys. So, get ahead men, get wearing those handbags. I guess it might even take my little ole clay pipe and baccy too. Now here's some music that's kind right for right now -- Paper Doll."

Walton switched it off. He felt sick again. Suddenly he yearned for the peace of Indonesia. He remembered his dream. That golden beach. Lestari in that bikini -- and out of it. And Nani's cuddles.

That was it. He'd call Lestari, just to see how things were of course, to see if there had been any news. But, he guessed he ought to call Grace first. He'd been a pig with her this morning. It was just the wrong moment. He'd apologise and be nice to her. She'd be OK.

Somewhat frostily Joe Simon's secretary connected Grace to the Vice President. What was it this girl had? Simon too was busy. Grace could tell from the clipped tone of his voice. It struck her that the last man to treat her properly was Bob Powell -- and just like the rest he was using her; at least she guessed he was.

Quickly she relayed Walton's message that he needed to stay on in the field. Was that alright? Simon said it was and asked about information for the press office. Had Walton said anything? "Nothing exactly printable. He did suggest some reporters should go out there and see for themselves, but he didn't know then it was a restricted area."

Joe coughed apologetically. "Nor did I until just now. I have to ask the President about that. Incidentally when you next speak to Dick tell him he might have to come back, like it or not, so we can fully brief the President. I think he's got some ideas and might need Dick's advice. OK?"

"Yes, sir".

"Oh, and Grace. You sound kinda down and it's none of my business, but if you had the time then Ruth and the boys would love to see you one evening. Just call her anytime, she'd love the company." He rang off.

Grace tidied her desk and drove home. On the car radio the bulletins were still full of the PWN news. It was now the nation's number one story. Powell was getting full credit for breaking it and now there was reaction from around the world. The air was full of pundits. Defense men, forestry scientists, financial forecasters, the whole world was getting in on the act. Reports of defoliation were now starting to come in from all over the Northwest. Acid rain was spotting cars and the paintwork of houses. Was it connected, the reporters wondered?

The White House denials majored on spelling out that acid rain was nothing new. There was no possible connection with PWN. How could there be? As for the PWN itself the White House acknowledged there might be a local problem, but quoted scientific sources proving that PWN had never survived before that far north.

Back at her flat the pictures on Grace's TV showed the evidence. Bare, stark trees against the sky.

Later that night Walton called her. He was calmer now, apologising for his reaction of the morning and explaining what he had seen. Could she understand? She could. She told him of the press reaction in Washington. Walton confirmed it was much the same in Oregon though there -- because of its past pioneering on the ecology front -- there was more bitterness.

Grace told him of Joe Simon's hint that Walton might be recalled to advise the President. He was flattered. Should he come back straight away?

"It's up to you -- but you told me this morning you were too busy." She sounded slightly put out. Walton took the point, but if the President ordered him back there was nothing he could do about it.

"Don't you want me back soon?" he asked.

Grace was surprised. She hesitated. "Of course, I do you dope, but you could have told me it was me you wanted to see and not the President." They said goodbye and Grace went to bed.

Walton called Lestari. At least she wouldn't hassle him.

## CHAPTER FIVE

Walton blinked, rubbed his eyes and stared once more into the microscope. Was he dreaming again? He wasn't. He was looking at a pine wilt nematode -- and it had moved. He'd brought it back with him from the Oregon crag that morning to double check that it was PWN. It was the classic textbook diagram brought to life. Just one millimetre of cylindrical worm. The scientist in him was fascinated.

Walton straightened up and walked over to the window, looking out at the giant trees in the distance. He wondered how much longer they'd be there. Could this David sized worm kill those Goliaths? He moved back to another microscope, this time studying the beetle he'd collected with the PWN. There was no doubt. It was a beetle in the family of a strain known as Monochamus alternatus. It was that beetle which had cause trouble in Japan. The worm became a parasite of the beetle as it emerged from trees and went into others. So, from the depths of space the nematodes had floated back down through the stratosphere and into the forests, homing in by some weird survival instinct onto the one insect which would guarantee their survival. So, they would live, and the trees would die. From previous scientific papers Walton knew they would now breed and spread at a phenomenal rate.

In a mere matter of months in normal circumstances they would kill thousands of trees. With acid opening up wounds in the trees they'd kill even faster. At least here, in this location, they would. He wondered about the forests of Indonesia. And, he mused, would Canada be safe?

Dozens of times now he'd gazed at the sparse information in the textbooks about the nematode and its relationship with the beetle. Now he went back to it again. The passage began almost apologetically: 'There is even less information on nematode pathogens of pines when grown as tropical exotics. However, this is simply due to lack of study by the appropriate experts and should not imply the problems don't exist.' It went on to describe the process by which the disease was spread.

It ended 'There is no doubt that the continued extension of exotic pine plantations in the tropics and the South will be followed by the appearance of new diseases caused by nematodes.'

Walton put down the book. By now he almost knew the passage by heart. It never got more cheerful. He wondered again about pesticides. But they couldn't be the long-term answer. Long experience had taught him it was only a matter of a few months or maybe years. Walton cursed aloud. The scientific assistant, silent until now, asked if anything were wrong. Walton grinned wryly.

"If you can think of a way of managing without wood and paper and all that we get from them then there's nothing wrong at all."

Walton went back to the first microscope and his study of the nematode. It was, he noticed, somewhat larger than the textbook model, maybe a third longer and certainly it looked slightly bulkier. As nematodes went, he mused, it looked like the healthiest specimen he could imagine. Maybe it was something to do with its birth in space.

Very carefully, he cut along its back, then compared a section with that of an illustration. It seemed to have a tougher skin than the classic model. Almost like an overcoat. With that thought his last hope vanished. Until now there had been only one possible salvation for the trees in North America if the PWN landed there alive. The hope was that they'd be killed off in the cold of winter.

Now, again perhaps due to their months in the cold of space, it looked as though they'd adapted to that too -- growing a thicker covering of skin to protect themselves. The textbook phrase about not being known under 20 degrees Centigrade would not then apply.

Once again, man had interfered and in so doing had re-written nature's manual. Back at the window he stared at the trees, rising away from him on a slope. He visualised that slope without trees, how the rain would fall, form small channels for itself and start off the process of carrying away silt.

Soon the soil would be gone, swept away to block some river or dam somewhere; soil that could have grown food, kept people alive, particularly in Indonesia. It was hopeless, it was inevitable -- and it had all been done in the name of scientific progress.

Involuntarily he shivered. For the first time, he felt scared. God help America. He checked his watch, worked out the time in Washington and called Joe Simon.

Joe was halfway through breakfast, the boys at their most boisterous, Ruth fussing them along. He took the call, ignoring Ruth's pleas for him to finish his breakfast while she handled it. Briefly Walton broke the news that the PWN was alive and distinctly healthy. It was, he reasoned, well established in the trees now -- the specimen he'd studied had come from the inside of a tree that was already dying. Because of their slightly larger size the PWN worked that much faster. What normally took months could now start to happen in a matter of weeks. It was going to be a remarkably quick death by forestry time-scales. In any case, once diseased, the wood would be useless for export in case it infected foreign countries. The whole of Indonesia's economy would collapse in a matter of months -- and America's might follow suit as well.

Joe's mind was racing ahead, computing political repercussions, assessing the cost of aid programmes, mentally wincing at the thought of reaction from both the Russians and third world countries. But hadn't he read somewhere that PWN only worked in warm countries? Walton explained the way in which the PWN had adapted itself, reminded Joe that California wasn't cold and that in summer both Washington and Oregon could easily reach 20 degrees.

Joe blanched. Oregon --his beloved home of pine forests. "My God it's bad enough for the people in Indonesia -- and they're starting off poor -- but the world will help them. You think they'll do that for America?"

He pulled himself up short. It was time to be practical. They said their goodbyes, Joe saying he'd tell Lean and that Walton should return to Washington as soon as possible. After that he'd have to check progress in Indonesia.

Walton rang Grace; got her to book him a flight and set off for the airport. It was a brief boss-secretary conversation. "I'll meet you at the flat" said Grace. "Fine" said Walton, his mind still half on work.

***

In the WWN news room Bob Powell had been given Casey, the science man, as an assistant. As the man with the inside track on the nation's major story he needed a hand. Leads were pouring in; amateur

Deep Throats were emerging from all over Capitol Hill. Only one sounded special, an embittered old five-star general called Abe Vaughan. He was an acolyte of Daley, the previous President and had shared with him his hard-line views.

Now, rejected by the liberal Lean administration, he was rumoured to be ready to talk. Not publicly, but, as the saying went "non-attributable."

Powell wondered about Vaughan's motives, decided he didn't understand the military mind and called him. Once again Powell decided to bluff his way through. "Listen" said Powell "I know that apart from the PWN -- the worms -- there was acid on that satellite. And that's what's taking the leaves off the trees. Right?"

Vaughan gulped, swallowing the loyalty of years, ditching the dictum of West Point and the Pentagon tradition of silence.

"Mr. Powell, you're a very shrewd man. I congratulate you on your information, but, you understand, I couldn't ever be quoted on this."

Powell grunted "I don't care about that -- I don't need to quote anyone. But we do know the defense department took over control of the space lab and you were in charge. Right?"

"My military history is no secret" said Vaughan "and if the record says the defense department controlled the spacelab well who am I to argue?"

Powell grunted again. No denials so far. He put it to him again. "So, both the PWN and the acid were on the spacelab?" Vaughan coughed quietly. Powell took it as confirmation. Powell went on.

"So, what we have to ask ourselves is why the acid was there."

Vaughan had to prompt him from there. "Acid's pretty nasty isn't it? The sort of thing people throw at their enemies."

"An offensive weapon?" asked Powell.

"Sorry son, I didn't say that. I just said acid was often used as a method of attack. It was just a general, or maybe a general's observation. Personally, I believe in defense. Always did, especially in a job like that."

"Defensive" repeated Powell "Like protecting something secret on a satellite..."

Vaughan said "You know what our society is like. If someone's about to mug you then maybe you stock up with a little defense in advance in case the worst happens."

Powell got the point. It was a defense system built into the system for future use. But to protect what? "You wouldn't protect worms would you general?"

"No, Mr. Powell, I wouldn't. We all have our hobbies. Personally, I'm into guns. Old ones that is, not these new-fangled laser things we hear about. If it were those things you'd have to be sure they worked before you installed them somewhere wouldn't you?

Powell quizzed again.

"So, if someone threatened to get hold of your best gun you'd defend yourself. Right?

"I sure would Mr. Powell, I sure would. It's my right to do so."

Powell had got enough. Vaughan could have denied anything he'd put to him. Instead, for some unknown reason of his own he'd given him enough hints to substantiate his story.

The acid had almost certainly been put there by a shuttle which Casey now confirmed visited the spacelab -- and it had been a defense against anyone finding out, much later on, that lasers had been put on board. It was a sensational story.

But, just as a check, just in case Vaughan was leading him astray he'd have to put it to the press office at the White House. Maybe not the laser bit yet -- he'd still have to check that further, but now, certainly, he could confirm the acid came from the spacelab. People had only wondered about it until now. This time he had all the proof he needed. He put his story to the White House press aide. The man laughed nervously. "I guess you don't want an answer to this in the next ten minutes".

"No" said Powell "Thirty will do -- we're running it then."

"I think maybe you'd be in some sort of hassle if you did" said the White House man.

"Oh yeah" jeered Powell. "And what would that be?"

"Like your station has an FCC licence that is the President's direct prerogative -- especially in times of a national emergency -- and he wouldn't be too happy about scaremongering."

"Any provision about telling the truth?" asked Powell.

The White House man wouldn't be thrown. "I don't think that's mentioned in the code. I'll call you back."

He didn't.

The next bulletin carried the full story.

***

Walton saw the billboards of the newspaper follow up as he stepped off the plane. As he drove home his car radio was full of it. He suddenly felt very weary.

Grace was waiting at the flat. She looked stunning and sounded warm. He knew that at the slightest signal they'd have been off into the bedroom. He just couldn't handle it now. She felt snubbed.

Walton poured himself a large one, slumped into a chair and explained what he'd seen in the forests. He was alternately angry and sad, but all the time bitter. Grace sympathised, hoping soon he'd come around, notice her, even ask how she was. Sure, she realised it was serious, yes it was a tragedy, but what about them? Walton carried on about man's stupidity. She gave up.

"Do you want me to stay?" she asked.

"Honey that's a crazy question. Of course, I do. You know that don't you?"

"A girl needs to be told you know. She can't just be taken for granted."

Walton groaned. "Yes, honey, I know. But first things first."

She blazed at that. "OK, so slap me in the face, tell me I come second -- oh boy you really know how to treat a lady don't you?"

He stood up, drew her close, kissed her and apologised. He knew if it had gone on much longer he'd have told her to get the hell out. Lestari would never have acted like this. What was it about American women anyway? Would they never learn?

He said none of it and led her to bed. As she undressed he had to acknowledge her body was superb. But he couldn't go through with it. Physically he could probably have handled it, but not all that emotional stuff, not when she'd harangued him like that. Maybe in the morning...

***

Walton stirred, stretched, blinked and registered that it was morning. His bed for a change. He slid out, went to the john and walked back into the bedroom. Grace, naked, face down into the pillow looked like something out of a men's magazine.

He felt better now. He stood there, looking, her mouth pouting, almost childlike. Except there was nothing childlike about what he had in mind. The thought sent him erect and, as it happened, Grace rolled over and smiled up at him.

"Are you trying to tell me something?"

For once he felt embarrassed and started to climb back into the bed. She stopped him, made him stand there while she quickly threw back the covers. She lay there, posing, smoothing her hands over her breasts, caressing her nipples with her finger tips. Her right hand moved to between her legs. Gently and slowly her hand moved, then her body with it.

"Now tell me you haven't missed me."

This time he got on to the bed. He kissed her, she moved against him, her hands lightly covering him. Gently he caressed her, tender now. Walton was enthralled. This was Grace? This was the prim looking Miss Efficiency?

She let him go, held his hand to her breast and whispered. "Take me now."

This time he took her almost savagely, impatient, turned on, urgent and demanding. Afterwards she was all tenderness, dominated as she'd wanted to be, satisfied and wondering.

"You have to be the best lover" she sighed.

Walton smiled "And for a so-called lady you are a great fuck as well." She stiffened slightly, withholding herself from him for a second.

"For me you know it's always lovemaking. Don't you feel that?"

"Of course, I do. But it can't be the same every time can it?" he smiled.

She sighed again. "Maybe. But sometimes I wonder. I thought last night that if you had to choose between me and the White House I'd have come second. Is that right?"

Walton shifted, took his arms from round her. "Honey, there's no comparison. It's just different."

She persisted. Walton groaned inwardly.

She carried on "Last night when you came home I could have taken my clothes off at the first sight of you. I couldn't wait to make love to you. All you wanted to do was to talk about the crisis. You were into your head so far I just didn't exist."

Walton sighed, almost out loud. Bloody women. He held her again, gently stroking her face. She liked that, he knew.

"Darling," he explained, using the word she liked "it's two different worlds. And I thought you'd be interested. OK, so I told you all the news. Don't you see that means you're a partner, not just a good lay."

It didn't work. Grace was determined now. Her pride was offended. Her voice had an edge to it. "It's not just that. At the drop of a hat you'll go off again to Indonesia. I've tried a dozen times to get you there and only once did you call me. You say you were working but look at that tan you got. Was that working?"

Now Dick felt guilty. And attacked. All he wanted to do was sleep. "Listen" he growled "I was working my ass off. As it happened I did call once, and you weren't here. Do I keep running to a phone every two minutes like a hen-pecked husband who's scared of being caught out? Is that what you want?"

Grace was fuming now. They even rowed over breakfast. "I feel used" she kept saying. Walton left as soon as he could, saying he had something to look up in a library. As he left he asked Grace to check flight times to Indonesia in a couple of days.

She gave a mock salute and answered, "Yes sir."

Walton made one last try to mollify her. At the door he went to hold her. She offered her cheek. "I'll see you in the office" she said.

When he'd left she gazed from the window. Her instinct told her this was never going to work. He'd never change and in the end it would all be very sad. She wasn't his problem. He was.

She went home and changed. Bob Powell had promised to call her and if he did then she'd tell Dick and he could take it or leave it. In many ways the two men were the same.......a driving force that overwhelmed anything that was a challenge. Yet, in Powell's case, there behind it was a dreamer, an idealist, a sort of Don Quixote always wanting to right wrongs, fight a cause and, mostly, for other people. And he had charm. Where Dick was abrasive Bob had a touch of sympatico that was hard to resist.

Resist? She pulled herself up sharply. What was there to resist? He hadn't even made a hint of a pass at her. And if he did? No, not while Dick was around, not while there was a chance that when this was all over, he'd recover from the excitement of this brush with power. It had changed him too -- he'd developed a ruthless hard edge.

Maybe it was the influence of the White House. Yet Bob was more closely exposed to it and had been for years -- maybe he'd just got used to it.

She'd heard his radio piece the day before and it was good. And accurate. At least now the whole world knew most of the story. The thought jerked her back to reality, she flew into the car and hustled through the traffic to the office.

***

The phone rang. Joe Simon's secretary, frosty as ever, silently questioning Grace's late arrival wanted her to pass on a message to "Mr. Walton." That evening he was required to be at a Presidential briefing session. It would be disguised as a social occasion, a first meeting of the new boys with the President. In fact, it was a review of the events of recent days. Grace listened politely.

"Thank you" she said. "Mr. Walton does know. I think Joe told him yesterday." She grinned mischievously and put down the phone.

When Walton arrived, she told him he had a date with the President. She gave him the details of flights to Indonesia and said she'd put in a provisional booking for the next night. Should she call the Indonesian Vice President and let him know? Walton, trying not to look concerned, said he'd make the call.

"Is there anything else?" he asked.

"Only that Bob Powell called and I'm meeting him for dinner. You don't mind do you? After all you'll have grander company won't you"

Walton heard the sarcasm and replied in kind. "At least I'll be working. Will you be?"

Grace wasn't to be beaten. "Yes, as it happens I will -- it's useful to know what he's up to. But you're right -- it won't be all work. He can talk about other things. And he makes me laugh."

Walton gave up, went to his own office and worked on some papers. Tonight, would be fascinating.

***

Most of the White House staff were leaving when Walton arrived. The invitations had stressed it was an informal get together, with a finger buffet and, maybe, a few drinks. A social event. Yet each guest knew differently.

As the buffet lay virtually untouched they each took their places around the long table, Lean at its head, Simon on his right hand. Walton was at the far end. He knew most of the faces from television. Lean did the introductions, a scientist here, a military guy there, a man from

NASA, an economist and a financier, together with the Secretaries for State, Defense, Foreign Affairs and the other main agencies of state. Lean explained that as a meeting it did not exist. There were to be no notes or minutes; it was purely so he could outline the situation that had come up; a situation he needed their initial advice on.

He filled them in on the details of the PWN, the pesticide, the acid, the large-scale defoliation now spreading like wildfire -- and as fast -- through the forests of North West America and Indonesia.

"We also have intelligence reports that Russia has been hit in parts and I'm sure we'll hear more on that later."

He continued, "I'm told by Joe Simon and Mr. Walton here that we have a real problem. I have to say the economic repercussions could be catastrophic. There is almost nothing that will stay unchanged."

He paused, took a swig of bourbon and went on. "There could even be changes in the world climate. We've all suspected, and there's now some pretty firm claims, that global warming is happening anyway -- without this situation adding to the troubles. This could be the trigger point of a disaster. One result could be an acceleration of the deglaciation of the ice caps like West Antarctica, a warming of the earth to the extent that the crops we eat could no longer exist. That's putting it at its worst, but, as President, I have to look at all options. We have to work with those possibilities in mind."

He paused again and looked round the table. "It depends right now on a can of worms. If they're alive we have a problem. What I want from you -- and I don't expect details at this time -- is a headline assessment of how it will affect the work of your agencies. And, once again, as you'll now appreciate, this is both contingent and secret. Any leak, any suspicion of one, could cause wholesale public panic."

He turned first to the economist. "What could we be in for?"

The man shook his head: "Hard to know where to start. One effect though is obvious. If we don't have timber we don't have paper which means we don't have dollar bills. If people don't have dollars they can't pay taxes and as a government we go broke. That's just for starters. Forget the paperless society and computers -- they eat it up. Wall street certainly couldn't operate without it and how businesses would get on without it I just don't know. If you don't have taxes as a government you can't employ people and you sure as hell can't make welfare payments."

Lean stopped him there. “See what I mean gentlemen. Now science...”

The professor was thoughtful. “I’d say a huge chunk of industry just stops. Dead. Timber products get into just about everything. OK, so a lot are imported; but, for example, around 40% of our pharmaceuticals come from trees. Hence, a possible health problem. Other industries? Well plastic starts off depending on timber in one way or another though you could maybe substitute oil. But that probably means you’d have to ration gasoline, at least in the early stages. Maybe more later on. The upside of that is that you’d probably have to cut car emissions anyway to stop increasing the carbon dioxide levels in the atmosphere. We don’t have enough trees as it is to absorb what we put into the atmosphere. Yes, I guess you’d have to ration gas. From there the whole thing snowballs. Obviously the construction industry goes early on and I couldn’t see the auto industry lasting much longer either.”

Lean interrupted again, bringing in the Secretary of State. He said “You worry about world climate. I worry about world power. If what the other guys say is right then we certainly couldn’t get involved in things like Gulf Wars anymore. Poor nations have no muscle, no voice, no influence and just think what would happen if the old hard-liners decided it wasn’t Mr. Lentov’s day any more. Even with him they are still a world power -- think of the fuss they kicked up about the Serbs in Bosnia. If we’ve already cost Indonesia their forest then they aren’t going to be happy. And since Japan depends on them for a lot of timber then we’ll get the blame for that too. And can we really afford any more trouble with the Japanese. Let’s face it forgiveness isn’t their strong point. I guess we’ll have a lot less friends than we had before, and we are going to have very little say in how the world goes on from this point forward.”

Lean stopped him, moving briskly on to defense. The general tried to lighten the mood a little. “Well I sure as hell would like to see an army fighting its wars without paper -- makes my life a whole lot easier. But, seriously the guys have to be paid and that’s problem number one. If we don’t have the money -- if we’re a poor nation -- then can we afford to keep troops abroad, say in Europe for instance. We’d have to think hard about our NATO commitment, but could we afford to not be there? And there are practical things of course. Like how do you keep records of servicing war planes? You can't take computers everywhere with you. And we'd have to have some sort of priority on plastic -- and just think of the gas we'd go through. Again, long term, we'd have to think of having a civil order role back home. Really it does bear thinking about."

Lean now turned to Walton. "you've seen the effects so far. Just how bad is it?"

Walton's throat felt parched. He was nervous. He coughed, then started. "If you think of a graveyard sir you've got it about right. I didn't know officially until now that the acid was from the satellite but knowing that, if the PWN is alive then we've unwittingly given if the ideal conditions in which to start killing trees. If the trees were healthy, those in the tropics would probably be OK. Having seen them, if the PWN comes down there it's basically wipe-out. As far as America is concerned the story is worse. Nearly all those plantations on the west side of the country are pines and, being one species, puts them more at risk. Being damaged by acid gives them no chance at all. I think we have to presume those forests stand very little chance."

The NASA man interrupted. "From the records we have sir -- and this maybe helps Mr. Walton -- I gather that the PWN was supposed to be kept under some sort of control in the capsules by a new sort of insecticide. Maybe that could be used?" Walton shook his head.

"Sir, people have been trying around fifty years now to find an answer to PWN and nobody's even got near it yet. With us of course it's never been a big priority -- we've had other diseases to worry about like Dutch elm and oak wilt and PWN has largely been a problem for the Japanese."

Walton confessed "without knowing more about it I couldn't give an opinion but looking at the worst of it I have an awful feeling that long exposure to an insecticide would only allow the PWN to build up a resistance to it."

The President put in a word. "Mr. Walton, as of right now you have what funds you need, what people you need, and you must make sure every effort goes into getting some way of killing that PWN invasion. OK?

Walton nodded. "Yes. sir."

Lean turned to Joe Simon "You've been in on this from the start. Any comment?" Simon loosened his tie, leaned forward and looked round at everybody. "We've heard some pretty gloomy stuff here tonight and while I'll agree it's bad I'd like to try to put it in some sort of perspective. You know what's going to happen? First off we're going to have people bitching about a shortage of paper. To them the greatest hardship they'll know first is they can't get things gift wrapped. Yet there are people out there who can't heat their food bowls to get a decent meal because we've taken their forests for wrapping paper and tissues. They'll scream because they can't have paper round their burgers. Don't they know some people say hamburgers cost whole forests? Those forests are also people's homes and we're putting cattle in there instead of trees that could sustain the lives of thousands of people. All this for hamburgers? They'll probably shout too because their favourite newspaper gets cut down in size -- they'll maybe lose the funnies. Do they know the true cost of those newspapers? And another thing -- you ever looked at the trash cans around your home? Man, they're overflowing with paper, with packaging, with cardboard - and with junk mail trying to get you to buy even more. A few facts. Each day every one of us throws away one-pound weight of paper. Not much maybe, but it builds up. It means each year we throw away -- wasting -- a forest the size of Delaware. We only re-cycle a quarter of it. In fact, we waste four times more than is used by all the countries of the third world put together."

He paused. "I know -- you've heard it all before a million times. But I just don't happen to think that because I'm an American I have a God given right to waste, through greed, what could keep another human being alive. Just maybe we'll have a hard time for a few years. If that is what it takes in the end to cure us of being the most wasteful, the greediest, the most selfish nation on earth then I for one will welcome it. OK, so we might have to use our cars less -- each of us walking a couple of miles a day would save billions in health care. So maybe we can't afford to eat as much, but do you realise that we throw away a quarter of the food we serve -- and we all know most of us eat too damned much anyway. And if along the way we get less greedy, regard ourselves less as the kings of the earth then who's to say that's a bad thing? So, while we may feel kinda down here tonight let's look at it another way. Let's go out if we have to, let's make people face the facts, let's try to make ourselves the better for it. God know the time's long overdue -- let's try to make this situation an opportunity."

He stopped, half grinned and turned to Lean.

"John. You shouldn't have asked me -- you mighta known what you'd get." Lean was quiet.

"Joe. I think you might just have told us something important. Maybe we do have lessons to learn."

## CHAPTER SIX

Next afternoon the White House issued a press statement. It confirmed there was acid aboard the satellite -- and blamed Daley for putting it there. It was, said the statement, part of a scientific experiment and a routine one at that. The PWN -- as announced at the time -- was put there at the request of the Japanese and everybody could now appreciate the need for such work. The statement said Lean had been assured by the Forestry service that the effects of the acid would wear off, and a new growth of leaves was confidently expected by the spring. As for the PWN there should be no great problem. On past evidence -- and it quoted the textbooks -- it didn't operate in temperatures below 20 degrees Centigrade. There would, of course be the most careful monitoring of the situation, so there was no need for panic.

Powell was in the new editor's office when the statement came through. "That's the biggest collection of half-truths I've ever seen" fumed Powell.

"But it's what they said, and you'll report it as a fact."

"But it's crap -- don't you see that? Don't you know that acid was a defense against hostile astronauts? There could have been some sort of secret weapon on board."

The news editor stopped him short. "You'll report that this is their statement. And you know why?"

"Tell me -- it had better be good" stormed Powell.

The news editor beckoned Powell to sit. "I'll tell you why. It's because I was reminded this afternoon who owns this station. It's a big corporation and they're into wood in a big way. Today you cost them a lot of bucks and they aren't happy about that. Now they're not saying they're interfering in editorial policy. They're just saying that maybe you'd like to broadcast the official version for once -- not your version of what you think could be the truth."

"But I'm right goddam you. I'm right."

The editor sighed. "Listen I know that. We both do. But as of now you're going to leave town for a few days while it cools off. I want you out on the road getting the reaction from outside Washington. We'll make it sound good. You start in Detroit. Goodnight."

***

Powell went back to his flat, poured himself a stiff one, stripped off and showered. Tonight, he could just get stinking drunk. Hell, he'd forgotten -- there was the date with Grace and moving out to Detroit meant breaking it. At least it was an excuse to call her and she hadn't been that far from his mind anyway.

She took a long time to answer the phone. She apologised. She'd been in the bath. He bit his tongue.

With anyone else it would have been the perfect chance for the sexy remark, to say hold it he'd be round. But this wasn't that sort of girl -- at least not for him, not yet anyway.

She was curious "I thought by this time you were always in the bars, picking up the gossip -- the big ear in happy hour?"

Powell laughed "I've just been retired from that -- I've been exiled, to Detroit." She was puzzled. He went on "Listen as I'm going to be out of town I'm going to have to break our date, but I wondered..."

"You wondered if I was doing anything right now" said Grace.

"Well I thought at least we could maybe have a drink."

He sounded depressed beneath the banter. Grace felt he was in more trouble than he was saying. Gone were the normal jokes, this was a man who needed a shoulder to cry on. But she too was tired -- it had been an anxious, long day. The news of the morning, plus the press statement that even spelled a bad time for her too. Dick never needed much excuse to be away and she was in need of company.

Powell caught her mood. "Say look -- this might sound kinda risky and pushy, but I can do a mean steak and I do have some champagne on ice. If you felt like it you could always drop by for a quick bite."

Grace laughed, more warmly now. "Bob -- you sure know the way to a girl's heart. I never suspected you had such talents. How could I refuse? Steak, champagne and gossip -- what more could I want?"

Powell put down the phone and eye-checked the flat. It was normally in a state of relative tidiness, but it did bear the hallmarks of a bachelor life. Moving quickly, he hurled things into cupboards, then started work in the kitchen.

Given the incentive, he liked entertaining and he did have flair. There were prawns, there was steak and, somewhere in the freezer, some strawberries.

Candles? He considered them very carefully, then rejected them. She might think he was trying to seduce her. With that in mind he rushed into the bedroom. Christ what a mess." He re-arranged the duvet, retrieved the pillow from the floor and cleaned the bedside ashtray. It would do.

In the traffic Grace was having second thoughts. What if Dick rang? And her clothes. The silk blouse was OK and the skirt was chic. But why had she deliberately chosen stockings. What was she trying to tell herself?

If it had been Dick she'd have known why she was wearing them. Was she now thinking of Bob the same way? Bob....was simpler to understand than Dick, there was a lot more loyalty there.

She had a lot of time for loyalty -- she set great store by it. But if loyalty was such a big thing, how come she was going to another man's flat for champagne?

She got out of the car, straightened her skirt and ran her hand down her thighs. Still firm, no fat on them yet. She shook her head and said to herself. "Grace my girl, you're a frustrated, horny spinster. You've started liking a sex life too much. Hold hard."

Two deep breaths in the lift, another outside his door and she was back in control. But was she too prompt? Shouldn't she have kept him waiting? Should she be here at all?

Powell opened the door and she now saw him differently. The shirt was casual, but expensive too. His hair, she noticed, still had the dampness of the shower on it. They said nothing, both blurted out a greeting simultaneously and then laughed. He led her into the living room.

The table was set -- or almost. He apologised that everything wasn't ready, and she offered to help him.

They busied themselves in the kitchen and soon they were eating. Bob, full of jokes now, was obviously pleased to see her and now, with the champagne inside her, she relaxed too. Yet, underneath it, Bob was strained. Soon he poured out the story of his row with his news editor, the objection from the station owners and his impending exile. It was confession time. Casually Bob asked about Dick -- when was he due home? This time he wasn't making it a professional question and Grace saw the difference.

He added "I guess you know I'd rather he stayed out there. He's a very lucky guy, but I do have to say -- though it's none of my business -- that one day he'll cause you a lot of heartbreak.

She smiled. "I'll call you then. That's a promise."

They both knew when to break the mood. "Now" said Grace "I want a quick tour, I'll help wash up and then I'm off back home."

Powell gave a mock bow and joked her round the flat. In the bedroom he swept his arm grandly to indicate the bed. He patted the pillow on the left-hand side and teased her.

"I'm a right-hand man -- tell you madam are you a left-hand girl?"

She curtsied. "Why lordy me, sir, I don't know what you mean. How can a girl tell without a chandelier in sight?"

"I'll have it installed in the morning."

At the door it was awkward -- as they knew it would be.

Powell solved it, just putting both hands on her shoulders and asking if he could call her from Detroit. She agreed, swiftly pecked him on the cheek and left -- rapidly. In the car she paused before driving off. She had only just escaped. Not from Bob, but herself.

***

The next day in the office confirmed her worst fears about the White House statement. It always had to be bad news before the media woke up. The threat had been there for decades and now, just because one piece of hardware had gone wrong, it was suddenly the nation's only concern. Lean was going to have to do something more about it. There was obviously no point now in trying to wash away the crisis with words. He'd now have to spell out the truth. After all, hadn't he campaigned on the slogan "Clean with Lean"?

It wasn't just the press either. The entire lunatic fringe was in full cry. The doom merchants were making hay.

Simon called Lean, making the points. Lean agreed. If nothing better happened in the next couple of days he'd go on TV and talk to the nation.

Reluctantly Simon had to admit a couple of days would hardly make things worse. They'd agreed they'd wait until Walton reported back more fully from Indonesia.

***

Walton was shattered. Straight from the plane he'd gone to the laboratory in Jakarta. It was a carbon copy of Oregon. The PWN, again a third longer than the textbook version, was alive, fat and thriving. Samples of bark pitted by the acid had shown a colossal infestation.

That morning, using an Army helicopter he'd gone back to Kalimantan, the landing site. The brown stain had spread now, wider and longer. Other islands too had the same stain. The pilot, who'd seen Vietnam at its worst in the days of napalm raids, remarked on the similarity. Walton agreed -- it was just like a war. He was dog tired.

Back at the hotel he called Lestari's father. She took the call. He was out of town until the next day. "Yes Dick, Nani's fine, but she's asleep now."

"Could I see you anyway?"

She teased him. "Dick. You should know better. It is most improper in this country to visit an unmarried lady alone in her home late at night. Unless you have an improper purpose in mind?"

"Lestari. I have to see you. It's important."

She recognised the gravity in his voice, and, in half an hour, he was with her. She offered him a drink, he declined and now she could see his drawn face.

They stood in the centre of the room. A picture of Nani brought a lump to his throat. Lestari moved closer, putting one hand to his cheek. "Tell me what it is. You are gravely troubled." Almost in tears now he gulped. "I went to the laboratory today. The PWN is going to live and the trees are going to die. All of them. And. And..."

He could hardly talk now for the constriction in his throat. "And Lestari, I'm sorry. So sorry for you, for Nani, for your father and all the people. Can you forgive us?"

Slowly, tenderly, she put her arms round him. He held her, burying his face in her shoulders, sobbing now in remorse. Gently she led him to the sofa and cradled him until the tears subsided. He apologised for his tears, ashamed of them now.

She too was moved. "Dick. I know it's bad. For us and for you and for everybody. But it's not your fault. I know you care, and I know your tears are for us and for that I love you as a man who is gentle. You are the one who is trying to help... who else will do that for us?"

Dick was silent now, both his hands on hers. If only the circumstances were different.
She was so gentle. Instantly he felt guilty at even the thought. But she had mentioned love. And now -- for this moment at least -- he loved her. It would be so easy to say. So easy. God, he must snap out of this.

She didn't help. "You are such a gentle man. So full of love and yet you are ashamed of it. It shows all the time with Nani and you can weep for us too. Why do you hide it so?"

Almost gruffly, he pulled her close. "Lestari, one day maybe. One day. From now on my life will be murder. I shall be here for a week, then I'll have to go back. There's so much to do -- we have to try everything we can."

He sat her up to look at her face. "Can I just say this and no more. As a man I want you, have done since day one and I have to admit you're like nobody I've ever met before. But, I don't trust myself not to hurt you and I couldn't bear that. It's happened before and you mean too much to me. I could never make any promises."

Lestari smiled. "I think I am being rejected. But we have many proverbs here about time. There is always time and sometimes a time. Do you understand?"

She got up quickly -- as if feeling she'd said too much. "Now you must have a drink and then you must go. You can see my father tomorrow, but I must tell him your news. He will be angry of course, he is already, but it will not be with you. You see he too admires you and your concern. It is something you share."

He finished his drink, wondering whether he dared kiss her. She moved to him, kissed him quickly and almost jumped away. Walton fought for control, battling the urge to hold her. He knew then that he couldn't have stopped, and he had a feeling Lestari wouldn't have wanted him to. He left, his head reeling, and sleep came late in the night.

Grace took his call in the office next day. Both felt uneasy. With little to say she connected him to Joe Simon. Walton told him the worst, that the PWN was spreading fast. Nothing they could do to stop it. The whole area had been showered with disease. "What does it look like?" asked Simon.

Walton thought. "Like God has crapped all over it. I guess he feels like we do."

***

In Detroit Powell was driving up Woodward, heading out towards Pontiac. Behind him the Renaissance Center towered into and vanished through the morning mist. He'd booked in downtown, rented his car from the Hertz office across the street, dodging his way over the tram tracks past Cobo Hall.

The whole city looked even more depressed than usual. The news of the close down of the National Car Corporation had almost brought riots to the streets. The UAW, setting out from its headquarters on the waterfront, had held a massive march through the city pausing almost in derision at the Renaissance Center. Built with Ford money, when free enterprise went hand in hand with gas guzzlers, now it was almost a towering mockery. Now it funded itself on tourists flocking into the souvenir shops, wondering at the dizzying escalators, gawping at Canada from the revolving restaurant at the top.

The UAW had marched, singing their anthem "Solidarity for ever", and in the basement studio they'd worked overtime churning out the anti-Lean commercials for radio stations. They'd even paraded with an old classic car, deliberately vandalised. Graffiti style it bore the slogan *'The Lean Machine. Be Mean'.* And in the streets, the crowds of now idled car workers had stoned it into a windowless hulk. All the bitterness borne and endured from oil crisis to oil crisis had shown. For they'd been the victims. It had cost them their jobs. Now, once proud craftsmen were in the welfare lines. OK, so the union benefit scheme had lasted a full six months this time, but nobody doled out pride. Now, thought Powell, they might have other worries. He wasn't sure, but he had a feeling that if they didn't have wood, they didn't have paper and he was sure that got into cars somewhere.

He'd look it up sometime. He glanced up left to snatch a glimpse of the huge Fisher building, topped by its General Motors logo. Almost parrot fashion, he repeated to himself, "What's good for General Motors is good for America". And conversely he thought... Could they really end up in trouble? Maybe not. Maybe they'd just carry on switching their plants to other, cheaper countries and God knows there were enough of those on their shopping list. It was like a separate empire. Maybe they'd see it through. But would Ford?

At State Fair he made a right, down past the fairground to the school. On his left, rusting now, was the graveyard of the guzzlers.

In the end they'd tried giving them away -- and nobody wanted to know. He stopped short. This wasn't what he was here to do -- though it was part of the Detroit saga. The newspapers were screaming about the possibility of a crisis to come, and for sure it was on the radio and CNN all the time now. He rang the office at lunchtime, gave them a piece attributing most of it to *Automotive News*, bible of the car industry, which had done its usual excellent background feature.

It wasn't, he added, Detroit's only concern. Now the acid had hit the next state and was moving East on the prevailing wind -- heading for Washington. And, as if that wasn't bad enough, there was the concern of the chemical companies whose factories fringed the freeway near the airport. Detroit, he reported, wasn't just cars. It was chemicals too and only now were the companies beginning to realise how they could be crippled by shortages of timber-based products. It was mostly speculative but, he reckoned, it would turn into fact -- fast. He heard the bulletin on the car radio -- expecting to hear his piece first. It wasn't. Even though it was a home story, concentrating on one of the nation's most crucial industries the first item was about Indonesia. The situation there was now desperate. Whole islands had been evacuated. Suddenly it was a world story.

Joe Simon heard it too. He called Lean, urging him to issue a further statement, maybe even talk to the nation himself.

Lean said little on the phone. "Joe, can you drop by. There's some more news."

Simon went in straight away. Lean's hand was up to his left jaw again. He turned from the window as Joe entered. He sounded almost scared.

"Joe, I've just had Lentov on the phone. It's hit Russia too."

"Oh boy, that's all we need" whistled Joe. "How's he taking it?"

Lean grimaced. Joe shook his head. "bad huh?"

Lean savoured the suspense. "Not bad, just a goddamned shock."

"In what way -- John tell me please?"

Lean poured himself a bourbon. "Joe you won't believe this, but he knew all about the acid -- and you know how? It had worked exactly according to Vaughan's plans. They'd gone up there and the acid had killed one of their astronauts.

And just like Daley had worked out they hadn't been able to say a darned thing 'cos they couldn't let people know they'd been messing with our satellite."

"Yow" said Joe "maybe I will have that drink."

"There's more" said Lean "He knew about the PWN in Indonesia - they must have agents, and damned smart, because one of the Russian offshore islands caught it too. So, he knew almost for sure that we're going to get it too and that in the end we're going to end up in deep trouble."

"In other words, he's got the upper hand on us."

"Yeah, but that guy really thinks ahead. We did a deal Joe. He's promised to keep out of our hair and not make a big deal out of it -- but on one condition."

"I daren't ask" said Joe.

"Nope. Just that we don't get into bed with China."

"You happy with that John?"

"Joe, I have to be. I don't have any option. It won't be confirmed or anything like that. Nobody's going public on it but that's the deal. Main point is we don't get any hassle from Moscow. We're going to have enough to do without worrying about them."

"You know it's blackmail don't you John"?

"Yeah. I know. Oh, and one other thing -- I said we'd keep off his back on an official basis about dissidents. I know that's one of your beefs, but right then he had all the cards and he'd heard enough to play them if he has to. Remember what he could say about bugs and acid defense systems in space. Just one speech and he'd swing a dozen countries. And we're going to need all the friends we can get."

"But John -- that's immoral. How can we condone what's going on over there? How could you?" Joe was momentarily angry. "He turned on Lean "John -- there's a touch of Judas in that you know. That goes hard with me. I'll have to think about that -- real hard."

Lean frowned for a second. He couldn't afford to lose Joe now. How'd he explain it?

"Joe. I'm sorry. But he had me on the ropes. And remember it was only promised on an official basis. It doesn't affect private protest. Best I could do." And don't forget it's not nearly as bad as it was before Lentov arrived."

Joe couldn't argue.

Lean changed the subject. "The question now is what do we do. I think you're right in that I've got to talk to the country. You seen the news?"

Joe nodded. All day the agencies had been churning out stories about violent fluctuations in Wall Street, some corporations had lost a fortune, many of them were firms which had backed Lean's nomination. The dollar had slid badly at one stage.

Even the yen dived as experts realised Japan had no timber of its own and relied instead on Indonesia. There had been acres of bad news.

Lean looked up. "I've already done a rough outline of the speech. It's a bastard to write. We've got to spell it out, but we can't have panic. Some way we have to be positive as well. Have you got any ideas?"

Joe shrugged. "Tell 'em to pray?"

He corrected himself, realising it was no time to be flippant. He started again.

"You need some sort of context. Don't forget man's been screwing up the world from day one. The first fire changed things. The first animal he killed exploited nature. And look long term too. Twenty years' time we'll get the trees back. It's just a bad moment in time."

Lean, despite himself, had to chuckle. "Joe. You just aren't a politician at heart are you? You must be the only guy here who thinks beyond the next election."

He paused. "But you could be right. Maybe we should look on it as some sort of challenge?"

He got up. "OK Joe, leave it with me. I'll work on it this afternoon then show you the final draft later on."

Lean paused his pacing near the window, stopped and looked at Joe. "You don't happen to have a quote that fits do you? A sort of text or something?"

Joe smiled, though there was sadness there too. "I looked one up, but I just hoped it wouldn't fit. It's this *'Our land, compared with what it was, is like the skeleton of a body wasted by disease. The plump, soft parts have vanished and all that remains is the bare carcass.'* Kinda cheerful isn't it."

Lean had nothing to say, he was looking out of the window again. Joe left, he had a machine to mind.

***

The TV crew moved in at 7:20, ready to record at 8. The three majors and CNN had cleared space from 9 onwards. Lean thought he'd need about twenty minutes. A programme planner complained to the White House. Couldn't he make it fifteen minutes?

There was a big ball game that night. They'd paid a lot of money for rights, had all sorts of deals in place with sponsors and advertisers.

The White House aide put it gently. "maybe you won't have that worry much longer. Baseball bats are made from wood you know."

The planner whistled. "Gee that's bad -- hey this really is serious isn't it?"

The aide agreed and couldn't resist adding. "you know something else?"

"What's that?" asked the planner.

"Two thirds of the world still use it for cooking food. In South America three quarters of them don't have any other fuel. Goodnight."

***

At 7 o'clock, Joe looked through the final draft. It was fine. Despite his laconic air Lean had a firm grasp of the problem, had set it in context and, at the same time, retained a positive note.

He called Ruth from his office. "Honey. I think the boys should watch John's speech tonight. He's got a lot to say. I'd be interested in what they think of it."

It was now near 8:00. In the President's office Joe picked his way carefully through the jungle of cable littering the floor. A make-up girl was dabbing Lean's brow. Otherwise he looked calm -- the way he was always determined to appear.

Only one gesture betrayed him, and Joe recognised it. Lean was stroking his chin. Joe knew Lean was nervous. He had good reason to be.

The buzz of chat died down, the producer asked Lean if he was ready. Lean smiled at him, not seeing him through the glare of the lights. He even had time for a joke, promising them all a drink afterwards.

Joe shook his head in disbelief. Here was this guy, about to tell one of the greediest nations in history they were about to lose their standard of living and he could still seem that relaxed.

Lean stroked his chin once more, smiled towards the crew and said, "Let's go for it."

"4,3,2,1, Action."

Lean glanced up to the camera with an air of studied calm. "Fellow Americans. For several days and nights now, I have to tell you I've been on my knees hoping I would never have to make this speech. However, when I was elected by you to this great office I made you a pledge. A promise that is engraved on my heart. It was simple. National security permitting, I would always tell you, my partners, the whole truth. You may remember the slogans about open government. They weren't just slogans. They were pledges, solemn undertakings. Now, tonight, is a time to honor that pledge."

He paused, moved to stroke his chin, but withdrew his hand to the desk. He went on "I have to tell you that today we're facing what could be, perhaps, the greatest threat to our existence that we've come across in our short history as a nation. It is nothing to do with hostile powers but concerns an accident which was quite unavoidable."

Briefly, he outlined the official version of what had happened on board the satellite, the descent first of the acid, then of the PWN. He didn't say why the acid was on board.

He paused again. "It would be too easy for me tonight to apportion blame to others. I'm not here to do that though I have to say there are those, once here but now outside this building who bear responsibility for the initial actions. My duty tonight is to spell out the threat, what has to be done and, above all, to ensure you get the truth. In the course of the next few days I shall be outlining, together with my advisors a series of emergency measures and powers I feel necessary to help cope with the situation. I also have to tell you how serious matters are. You may think that a few worms in an isolated forest is none of your concern. How could it affect you? I have to tell you there is not one person in this land who will escape the effects, nobody who will avoid personal hardship. There is no way in which it can be avoided. No way at all. None." He sipped from a glass of water.

"But let's put it in context. Since the dawn of time man has in some way or another been exploiting the environment. Even if we were not facing this particular crisis we would still be in trouble. Scientists all over the world, for decades now, have been warning us that there is a finite limit to our resources. Only in the field of oil perhaps has that message been brought home to us. What we chose to ignore -- and I count myself in with you on this -- is that at the rate at which we were currently using wood we could well have run out in the very near future -- by the end of the century anyway. I ignored it. So, did we all. Now at moment, that truth has galloped towards us. The moment is now -- or at least only a matter of a few short weeks away. The truth is -- and it's taken me by surprise too -- is that we rely on wood for an astonishing range of goods. Paper is the most obvious one and I'm sickened by the news that in America alone we throw away more paper than is used by the rest of the world combined. But it's not just paper. It's not just the obvious tissues, cigarettes and dollar bills. It's far wider than that. It affects almost every single aspect of our industrial output, the goods from which we gain our standard of living. It is no exaggeration to say we face, long term, the prospect of national poverty." He took another sip of water. Despite the TV lights there seemed to be a chill in the room.

Lean looked up again. "Before I spoke to you tonight I was given a quote from Plato." He paused "He was a great Greek thinker from centuries ago. Listen to it."

"Our land compared with what it was is like the skeleton of a body ravaged by disease. The plump, soft parts have vanished and all that remains is the bare carcass." He looked directly into the camera now. Grimmer.

"Since coming into this office there is one single truth that has become even more evident. This America of ours is still the greatest nation on earth. But that didn't happen by accident. It happened because our fathers and our grandfathers -- and their women too -- made it that way. They worked, they had know-how. That spirit is still there. But we are still a nation in its youth. Our America is still a youngster. And, as all you parents know, youngsters can panic, they're a little light on self-restraint. We cannot as a nation afford to panic. I have to say that the measures I will be forced to introduce will include severe penalties for anyone found to be indulging in any form of exploitation or selfishness whatsoever. It is not a time for greed. It is not a profitunity. Instead it's a time for the reverse, when all our best qualities must come through. We have pride as a nation, let us all take pains to work together for a way of getting this licked. I don't know how this will be done. I have no magic wand. However, nobody in this administration will rest until there is some hope of either cracking the problem or finding ways to cope with it. I can't do it without your help and maybe that of our various Gods."

He paused again. "I gave you, just now, a quote from Plato. It was perhaps about despair, an obituary almost. I regard that however as defeatist. I don't think this nation is beat yet. With your help and those of all our Gods it never will be."

"I started tonight by saying I was honouring a pledge. That pledge was to tell you the truth. That I have now done. I shall continue to do it though that truth may often be painful. I could only tell the truth to a great nation, the greatest country, the greatest people in the world. What we have to do now is prove that." He paused, just for a second, looked down, then up.

"My fellow Americans. I have faith in you."

The lights dimmed. It was silent. After a full fifteen seconds the director quietly walked forward to the President's desk. "Do you want that replayed Mr. President?"

Lean shook his head in silence, his hands clasped in front of him. He unclasped them, put them together as fists and quietly brought both to his mouth. A second more and he straightened up, took a deep breath and smiled to an aide.

"I guess we could all use a drink huh?"

## CHAPTER SEVEN

Walton saw the speech in his hotel room. He poured himself another drink. Then he called Lestari. She'd seen it too -- and guessed how he felt. There was no point in talking about it. Instead, she tried to divert his attention.

"Dick -- there is a change of plan for tomorrow -- we have to use a seaplane. I'm coming with you and you must bring some swimming clothes."

Walton was floored. "But I shall be working. There's a lot to do."

She was persistent. "You said you wanted to look at one of the smaller islands. They don't have air-strips so it has to be a seaplane. Nani loves flying, she loves the beach and she's never stopped talking about you. Would it be possible?"

He laughed. "Put like that I can hardly say 'no' can I? Just as long as I can do some work."

She assured him "It will just be a break in the day. You have looked very tired. It will help you."

***

She was right. Taking a small inflatable from the plane they landed on a perfect beach. To any casual observer it would have looked like a family picnic. Nani squealed with delight at the water's edge, while Walton stripped off to sunbathe.

Lestari astonished him. Standing up, she slowly slid off her clothes revealing a stark white bikini beneath it. It was like a dream come true. He had to turn on to his stomach, his erection would have embarrassed them both. But he could not disguise his surprise.

"You like it?" Lestari asked, a trifle shyly. "Yes, I love it, but it's a long way from traditional dress isn't it?"

She chuckled "Yes, perhaps my father would not approve. But I am a woman...." Walton was acutely aware of that -- too much so for comfort. Hurriedly he switched his mind to pine wilt nematodes, guaranteed, he'd found, to dispel erections -- and sprinted quickly into the water. Nani splashed to him. He picked her up, pretended to drop her into the waves and she giggled in delight. Lestari looked wistful. It was good to see them. If only her husband could see Nani now.

Her husband.... guiltily, she realised it was the first time she'd thought of him for several days. When he'd first died it had been worse. Even when the first pain had abated, and the sudden tears had gone away there was still an emptiness that was hard to handle.

A shower of water stopped her dream. Walton and Nani were telling her to come back to earth. Walton looked at her a little oddly and, later that evening, having their, by now, normal drink he teased her about it.

"It was nothing" she said "it was just that Nani, loves someone, a man, to play with. She likes you a lot. She'll miss you -- people can so easily become part of your life can't they."

Walton just nodded. He looked her full in the eyes. "Yes, they can. I know that too."

***

Powell's editor had a question -- how fast could he get back?

"What happened?" asked Powell "I thought our dear owners wanted me out of the way?"

The editor explained. "They ran out of wood -- what was the point if there wasn't going to be any? So, get back here fast."

Powell was back that afternoon, revelling in the chaos that followed Lean's speech. Mostly it was financial. Wall Street had reacted first. Share prices yo-yoed as the brains of commerce worked on the likely effects. By and large the multi-nationals seemed to be riding the storm better at this stage. The pundits argued that as America ran short of some goods the giant corporations with tentacles abroad would be better placed than those at home. Some second thoughts then dented that first reaction. How would the domestic consumer market stand up if there were large scale unemployment -- and that now seemed inevitable, particularly in the auto industry.

Then, as the world-wide foreign exchanges opened up, the dollar started to slide.

Lean's brief mention of forthcoming poverty did nothing to help. By noon the dollar was at an all-time low against almost everything, bar the yen.

"Why the yen?" asked the news editor. "Because they get around three quarters of their timber from Indonesia -- they'll be almost as badly affected as we are" said Powell.

He gasped "But look at the Swedish kroner!" Like most of the European currencies it was soaring. Any country self-sufficient in timber was riding high -- and would for years to come. The whole balance of world trade was swinging round. But there would be third, fourth and fifth thoughts, deals would be made and, he guessed, fortunes too. There always had to be someone who'd make a quick buck somewhere.

The domestic news was just as intriguing. In the treasury building there's been an all-night meeting to gauge the effects on the administration. No paper meant no dollar bills -- and how were people going to pay their taxes?

For all the talk of a cashless society it was still a long way off yet. Bert Casey joined the throng at the screen. He had a wry observation too. "Just as well we don't have to rely on agency tapes any more. They'd soon run out of paper."

Around the country people were trying to second guess the President's emergency measures. For all his talk of pulling together there was a distinct dash of personal survival coming through.

Lean had talked of God -- and some people had gone to church. Like in Atlantic City where an armed gang had gone in during the night and made off with all the pews, the bibles and the wooden plaques on the walls. In Georgia another enterprising gang had broken into a funeral parlour and made off with the coffins -- then gone back for the doors as well.

Even Hollywood was panicking. Technicians there had realised the raw material -- cellulose -- derived from wood was in danger of destruction and that stocks of film would have to be conserved.

Powell walked away deep in thought. The story, for once, was just too big. Where on earth did he start? He now had the job of running what they were calling the paper desk -- though, thank goodness, it was made of metal. Idly, he wondered what audio tape derived from. He wouldn't be surprised if that didn't come from wood somewhere along the way.

He stubbed out his cigarette -- and glared at it. How much longer would they have paper for those? No use switching to a pipe, he mused.

But then there was always clay, though wasn't wood used in ceramics too? Cigars maybe? But, then how would women fare? In his mind's eye he could see the ads now -- "Get sexy with a cigar." But they wouldn't be on billboards. Maybe billboards would have to be re-cycled. Just think how clear the roadsides would be – at last.

His mind was running amok and he had no doubts that around the country, the world maybe, there were others doing the same. But these were trivialities -- like Casey's concern about not having toilet rolls, like newspapers worrying about their own fat fate if and when newsprint rationing was introduced. The word was, that was due to be announced first.

Powell wondered what would come after that. He had to take it logically. America was in trouble. What was it Lean had said? Facing poverty. In this world you get muscle from money. Without money America couldn't afford to keep or even send its troops abroad. No more excursions into the Middle East. The whole balance of world power could change. He grinned to himself -- he bet nobody else had thought of that yet.

He yelled at the political man. Get me a general. Fast." He shouted Vaughan's number at him. "Ask him if we're that poor if we can afford to stay in NATO? Could we be involved ever again in something like Bosnia, the Gulf War, Iraq whatever?"

Now he could write the bulletin. Whatever Vaughan replied he could pose the question. That should keep him ahead of the other stations 'til the main evening show.

The news editor thought it was a bit wild. He'd rather have dealt in established facts. "It's news Bob. You gotta report facts, deal with what's actually happening -- you can't go dreaming up theories."

"It's not that far -fetched" retorted Powell. "Just think about it for a couple of minutes. I'd say it was more probable than possible and I bet the Pentagon would too."

The news editor wasn't beaten. "But it hasn't happened yet Bob. So, there's nothing to report."

Powell grinned disarmingly. "There is -- it's just a fact that hasn't been reported yet." Grudgingly the news editor gave way. In his bones he sensed trouble.

It did get reported. By mid-afternoon, after the announcements about newsprint being rationed and gift wrapping being banned there was one small, sinister, passage. "In view of the national emergency the President has re-activated the special powers sub-committee of the National Security Council".

"What's that?" asked Casey and Powell together. The editor winced. "The first casualty."

"What do you mean?" asked Powell.

"It's a relic of the second world war. It's a committee which, to all intents and purposes, monitors the output of the media. It's not a censor as such, but there is a duty on us to check with it when we feel we have something sensitive to national security. Like whether we stay in NATO for example?"

"So, what can they do about it?" asked Powell.

"Simple" said the news editor "if they don't like you and what you're doing they just take you off the air or the streets or the screen. It's that easy. There's no appeal and there's also an understanding we all have with the White House that there's no public reference to that committee's media function."

"So, it's censorship. A gag. That's unconstitutional" fumed Powell.

"You're right, it's the first casualty again."

"They call it preserving national security and you know they could be right" said the news editor. "I don't like it. I hate it, but I sure as hell have to abide by it. And so, do you; don't ever forget it."

Casey chipped in. "What's this first casualty stuff?"

The news editor groaned. "Don't they teach you anything at those colleges? My boy it's just about one of the most famous quotes there is about war reporting. Remember it.

"In a war, the first casualty is truth."

"So, we at war boss?" asked Casey worriedly.

The news editor sighed. "If the White House says so then who's arguing?" He was called to a screen -- and burst out laughing. Powell and Casey looked baffled. It was the latest White House announcement. Though taxes weren't due for another three months there was, as of now, an immediate discount of five percent for those who wrote the IRS a check for the year right now. Debits or credits would be adjusted later.

"Why'd they do that?" asked Casey.

"Because" said Powell "it's so they can go on paying themselves and keep on handing out welfare cheques -- and pay the armed forces too I guess." His voice tailed off quietly there.

"Can we say that?" he asked the news editor.

The editor grinned." Depends how you say it. You can always include it in a review of where the money of loyal taxpayers will go. I reckon we could get away with that. At least, maybe, it could prove your point."

"Damn my point" said Powell "do you realise what it means? It means the whole damn country could go broke by this time next year. One thing's for sure, the welfare payments bill is going to go sky high -- and how do you pay them without dollar bills?"

***

The crime man shuffled by on the way to his desk; muttering quietly to himself. After years of whispered confidences and tips in the back corridors of police stations, he had difficulty talking at full volume.

"What is it Jim?" asked the news editor.

"The Mafia had it right. On Wall Street today they're making fortunes in just hours, gambling on commodities and futures. And that's legal. You know what just happened downtown? Some stupid kids hijacked a bus that only had the driver on board. It was one of the old sort that still gave out tickets. You know what they did?"

"Tell us Jim" said Powell.

"They robbed him of his tickets. They told him they were going to recycle them. I tell you that place went mad today."

"What else?" The news editor was terse.

Jim's weary face cracked into a grin. "You ain't heard nothing. Some of the cops who moonlight doing security jobs have just got themselves a new assignment. They're standing armed guard on churches, hired by the priests to protect the bibles. You ever seen old ladies being frisked for hymn sheets when they come out of church by a six-foot armed Mick? Don't bear thinking about." He adjourned to his hip flask.

The news editor growled after him. "Don't celebrate it. Write it damn you. I want it on the bulletin."

Jim nodded -- and took a swig anyway. "Sure, you can afford the paper? I could just do it off the top of my head."

"Shit" said the news editor "why didn't I think of that? Maybe you should just record and edit if you have to. No better -- do it live if you can stay sober long enough. I guess we'll be doing a lot of live stuff from now on."

He turned on Jim "Just how long is that bulletin anyway?" Powell checked. It was five minutes instead of the normal three. The news editor fumed "How come you're that much over?"

Jim replied quietly "There's a lot to tell boss. A lot. Ask them upstairs, maybe they could find a gap between the Kleenex commercials."

Despite himself the news editor grinned. "OK you've got a point. I wonder how many more commercials are missing? Hey, I wonder how that affects our income?"

"You think we could go bust too?"

"Not a chance" said Powell. "Last I heard they were filling up with government announcements, how to make your own paper was one. Then there's the stuff about how you should save it for your country. Save it... what a joke. If you listen to Jim half the country's out on the streets either buying it or stealing it. What a crazy place this is."

The bulletin went out. The night shift took over. Powell headed for a drink with Casey. In the lift was the junior detailed to find Walton. Powell had forgotten. So where was he?

The junior didn't have much of an answer. They said Walton was out of town somewhere. It could have been Oregon. Probably wouldn't be back for days.

Powell grinned. Every cloud had its silver lining.

***

In the White House, Lean was restless, itching to do something. Anything. The reaction to his speech had astonished him. All his appeals for patriotism, all his statesmanlike urging that people should work together, had seemingly fallen on deaf ears. Almost as one the American people had opted for personal survival and/or profit. What sort of nation was it that would tip dead bodies out on to the floor to get their hands on a coffin? Even the long years of cynical wheeling and dealing his way up through the system hadn't prepared him for this. Maybe it would all calm down after a while, perhaps people would see that only together could they survive.

But, for the moment, panic was rampant. It was the worst of the American nature which had surfaced first. Almost overnight. And some of the media coverage had been trivial beyond belief. Even at a quarter of the size, the newspapers still found room for debates about whether women would smoke cigars. Much more of that trivia and he'd slim them down even more. That would teach them.

Perhaps they needed some sort of example? Maybe he wasn't being enough of a leader. What would Kennedy or Clinton have done? Desperately he needed time to think. He'd always tried to sleep on a problem. Fat chance of that now.

Twice during the night European leaders had been on the air waves, worrying about rumours America might pull out of NATO. That bloody radio reporter. No wonder the media had to have restrictions on it. So, he'd had little chance for sleep, little time alone to debate the options. Yet here in his office it was too damned lonely. He'd been warned, but he'd never get used to it.

What was needed now was some sort of gesture, even if only for a little while, that would give people hope. People took strength from things like that. But what could he do? On his desk were forest service reports for the past years. Time and again they bore the same phrase. "Target not achieved due to low level of funding." Or, in another one "The accomplishments reflect the low level of government assistance."

Then something else caught his eye. A passage on forest service research. "Scientists have found that wildfire research is necessary and plays an important role in maintaining healthy forests." He didn't bother to read the rest. It was the general idea that caught his imagination. Fire. Maybe the PWN could be burned out. How could they survive it? The forest people were always wanting more money for firefighting, justifying bigger budgets on the grounds that it was cost effective, pointing out the havoc caused by forest fires. So why not used that weapon for once? After all, at this point, the trees were useless anyway.

He walked to the window. Yeah, the idea appealed to him. Politically it made sense. Fire was dramatics. He could use the air force to do the job. He could even go to the airfield to see them off. That way he could be seen to be doing something. And it might even take people's minds off things for a while. He phoned for Joe.

He was busy on a long-distance call.. It was with the President of a multi-national tobacco company. A worried man. And an angry one. "You realise how much money we could lose?"

Joe was patient. "You're not the only one you know. There are hundreds of others."

The man wasn't mollified. "You know how much we pay you in taxes? Don't we get anything back for all that? Shouldn't there be some government compensation?"

Joe nettled slightly. "You only pay on profits. You've still got cigars and pipes."

The man exploded. "We've been into all that. People want cigarettes. And all that clay pipe crap. There's some damn thing in the ceramic process that stops you making them without wood. And mister I tell you I asked my wife and there's no way she's smoking cigars. But it isn't just that -- there's jobs too. You think about that?"

Joe was sterner now. "This wasn't our doing. Right now, we're trying to sort it. Why not get paper from abroad?"

"With today's dollar? It don't buy nothing no more. There's no way the public would pay the price."

Joe's patience was running out. "Sir, I have to go. We all have problems. Goodbye."

The man was still shouting as he put down the 'phone. Joe had little sympathy. The world had to be better without cigarettes.

The phone went again. He groaned. Lean sounded chirpy. Joe couldn't think why. He grunted that he'd be in.

***

All day he'd dealt with people who thought their campaign funds bought them an office in the White House. It had been one long catalogue of greed. He'd not had one call from anyone offering to help anyone else. He just wanted to go home. Even last night, slipping into the synagogue for a quiet moment he'd been besieged. Even the rabbi had been changing the lock on the door.

Lean was on his back on the floor, doing stomach curls. He bounced up, still puffing slightly. "Do you keep in shape Joe. You should you know."

"I jog most days" said Joe, not mentioning it was a full three miles each morning.

"That's my boy. Now, listen -- I've had an idea. It may not work perfectly, but in a funny sort of way that's almost not the point. As I see it the nation right now needs a leader, a gesture from the top, needs to see we're trying real hard, trying anything to stop these damned worms. Get my drift?"

Joe sat down. He felt he needed to. "I think so -- I think the people are searching for something."

"Damned right they are Joe. And I'm going to give it to them. Know what I'm going to do?"

"No John".

"Joe. We're going to burn out those worms. Bomb them. Use the Air Force. Blast them to hell. They'll never stand fire."

"John that's crazy. You'll just burn dead trees. And we need the wood."

"Listen Joe. You know I made part of Oregon a restricted area. It was just to leave the Forest Service some space to work in. But, say we took a bit of Alaska, got everybody out of the way then we could bomb it just as an experiment; just to see if it gave us some answers."

"And if it doesn't?"

"Joe. That's not the point. We'll have been seen to try."

"I think you're nuts. But I do prefer you did it to Alaska not Oregon."

"Joe. What's the problem? We've got nothing to lose except some already dead trees. OK, so we need the wood, but we'll only use a small area. And it'll boost morale. We'll get maximum TV coverage -- I'll go to the base to see the pilots off. It's a shame they don't have bomb doors on Air Force One. They don't do they?"

By now he was almost bouncing around the room. "Say, maybe I could go up in the bombers, drop the first one. Is it bombs we need or is it napalm?"

Warily Joe interjected. "It's bombs. Napalm you remember takes the leaves off trees. We already did that bit."

Lean barely paused. "How do we get the animals out? We can't get accused of killing wildlife. Can we ask Walton about all this?"

Joe stood up "That's a good idea, ask him first. Maybe there's a snag. I'll call him."

Lean turned. "OK. Do that. But tell him I'm real keen to get on with it."

Joe left, shaking his head in disbelief. Whatever happened to sanity? Yet, in some crazy, twisted way maybe Lean was closer to the people. Maybe they did need a leader for the moment. He couldn't tell anymore.

***

In his office the phone was ringing. It was Lean again. "Hey Joe, I have a good idea about moving out the animals. I just remembered there's a thing called the Young Adult Conservation Corps. Perfect job for them. We've got the camps, we've got the people, they can just go in and drive the animals out. And if there's any animals left over well at least we made the effort. Say, why don't we just get started with that anyway?"

Joe said he'd call Walton first.

***

Walton was in the bath, sensuously lathering himself and wishing to hell Grace had come out with him. He could do with her right now. Maybe if she could come out just for a night or two....

He cursed at the phone. Joe told him the President's thoughts. Walton laughed outright.

"That's the best joke of the day. Hey that guy's really got a sense of humour. You mean he's serious? He can't be. Does he know what he's doing?" Joe sounded a bit puzzled.

Walton explained. "Joe you have to tell him. Smoke from a forest fire can travel over hundreds of miles. Now that hot air could lift up all sorts of cultures, spores, fungi, almost anything. We could just be taking away something that will help us fight the PWN. You can't let him do it."

"Dick, he's determined to. I know the man. We can't stop him now when he's got the idea. He knows it probably won't do anything constructive, but he needs the gesture."

"And Joe -- this Young Adult Conservation Corps idea is crazy. I've worked with them. They're mostly city kids. They'd get lost. They'd get attacked and some would just die out there. Doesn't he know how dangerous a hungry grizzly can be? You can't go in there unless you're armed, and you can't arm those kids."

Walton reached for his towel. "Listen Joe if he wants to set fire to one section then OK, let him. But why bombs? That way he's bound to kill animals. If he just set fire to the edge at the right place the wind and flames will do the rest for him. The animals -- well a lot of them -- will get away."

Joe listened patiently. Then, somewhat cynically he outlined Lean's plans for the big personal TV send off for the bombers.

Walton groaned.

Joe carried on "Dick I feel the same way, but in the end I back the guy. I just have a feeling that somehow it might strike the right chord with the people, give them a lead, maybe boost their morale in some crazy way."

Walton groaned again. "Joe -- what was the point of calling? The guy's going to do it anyway isn't he? But just do me a favour. Tell him just how wrong it could go. I'll watch it on TV."

They chatted briefly about other developments, Walton promising that he'd be back within a week.

***

"How are you?" asked Simon.

Walton laughed. "Dead, bushed, worn out. But, apart from that, fine. How's Grace by the way. I keep trying to reach her and can't get her in the evenings."

Joe was puzzled. "She seems fine. If I see her I'll get her to call you."

Walton headed for the bath. He was interrupted by the phone ringing again. It was Lestari. She was concerned that he was working so hard. Had he eaten? He realised he hadn't, but he wasn't that hungry anyway. He told her about Simon's call and Lean's latest lunatic idea.

"You sound depressed" said Lestari.

He agreed. Lestari was quiet for a moment, thinking. Then she said "Would you like to come here for a light meal. I can arrange a car."

"Will your father mind?" asked Walton.

She explained that he'd had a meeting with some senior advisors, but now they'd all left for the President's palace, taking some servants with them. They had talked about staying overnight.

She was in the garden when he arrived. The air was soft and all around them were the familiar noises of a jungle night. He held her arm, discreetly in case a guard was too curious. When they went in the food was ready. Lestari dismissed the servant, addressing him in their own language. Guiltily, Walton, wondered if he'd gone for the night.

Two brandies after the meal and he was feeling warm, though a trifle light-headed. They'd said almost nothing out loud, yet somehow the way Lestari was looking at him was different. They moved to the sofa. It was long, low and wide, yet she sat close to him.

He held her hand and studied her face. It was troubled. "What's wrong?"

"You'll be going soon -- back to your other world. You are important now. And busy."

Walton nodded "I'll probably have to go back tomorrow. It's almost certain now. But I shall be back here. I guess -- if your father wants me, that is -- I'll be back here a lot."

She smiled then. "Nani will be pleased."

"And you?" Walton asked.

She took his other hand. "Dick I think you already know how I feel about that." He straightened up, slipped his left arm round her shoulder and eased her to him.

"Lestari I told you before. As a man there isn't a part of me that's not aching to make love to you. And it's in my soul too. It may sound corny, but really I just haven't met anyone like you before. And, I as I told you before, I can't make any promises." His right hand was still holding her. She was silent. She moved it to her breast. There was no bra. The nipple was erect.

She turned, breathing more heavily now and kissed him. "I'm not asking for promises. And I am a woman too." She said nothing more, but took his hand, stood up and led him towards the bedroom. He caught her at the door, almost roughly turning her towards him.

"Lestari. Are you sure?"

She was still panting slightly. "I am very sure." Her clothes dropped off as if by magic. There was only a dress. Walton, halfway through fumbling down his shirt buttons stopped short. He gasped. He had never seen anything so alluring. Was it the brandy? Was he dreaming? Her body, lightly tanned, was so firm, the breasts so pertly uplifted.

She smiled, and stepped towards him, putting her hand to his mouth. She undid the remaining buttons, then fumbled with his belt. He was quicker at that. She knelt before him. He felt her fingers, then her lips. He gasped in ecstasy. She slid on to the bed and he followed. He was in a daze, but she was whispering to him. "Dick, for me it has been a long time. Do you understand?"

He grunted. She half raised herself over him. He stroked her breasts. Her hands slid down his stomach again. Now he found her -- wet and ready.

"Now" she said. "Now." He took her then, astonished by her passion. And then the world was a blur.

Afterwards she gently pulled a cover over them, folded his face into her breasts and nursed him. After all these years she was a woman again and she felt the flow. All the love, all the care was there to be shared again. Shyly she looked at his face. He was handsome too, she thought. And he'd been so gentle, so caring.

Walton felt like he'd died. It had been pure passion and love without a touch of artifice. It could never last like this. He'd end up as always, taking her for granted, seeing on her face the pain of his selfishness. In the end it had always been that way. But he was saying that now and realising too it was a reflex. Maybe if he just relaxed, didn't expect perfection things might work out. If only they could. If only."

She stirred now and left the bed. He dozed. She woke him with coffee, and a smile. She whispered "I wish you could stay. I would like to wake with you here."

Walton took the hint. He'd never had a servant problem before. He smiled at the thought. He dressed and they sat together as he had his coffee. He took her hand. "Lestari. I don't normally say things like this, but right now I love you. Can you make do with that?"

She smiled "But I also have memories now don't I?

The goodbye kiss was warm. Full now in their knowledge of each other's body. He walked to his car, still half in a daze. Silhouetted against the light of the house she looked like an etching. She was etched into his mind as well.

The army car swished its way back into town, Walton slumped in the back, deep in thought. He was still slightly shell-shocked. It hadn't been just an easy lay. She'd wanted him as much and he really cared about her.

Yet, wasn't it the same with Grace? It was, but there was in Grace, this demand to be treated like some treasured object. She was, but there was never any relaxation, it had him on edge a lot. He wasn't sure he was up to that. He wondered if other men felt the same.

Maybe, thought Walton, he was an idealist. Maybe he only ever wanted the cream and not yesterday's milk. But who could stand that all the time either? He really didn't know. All he was sure of was that when it came down to it he was a shit. Somehow the women loved him and yet he always ended up hurting them, even though he didn't want to.

Why couldn't they just relax about the whole thing? And he just couldn't stand the recriminations -- and they always came, no matter how plain things were made to start with.

Maybe Lestari's Oriental background would make it different. Could it be this was just fantasy, or perhaps they really did know how to make a man want to stay close. It was peculiar but he didn't feel he'd been unfaithful to either of them. He wondered if he'd care if Grace went to bed with Bob Powell. Perhaps that was the difference. If he was jealous about that then maybe it would be a sign he really wanted her. He gave up. He'd been trying to work it out for years and he'd always failed. What was to special about today?

Next day he flew back to Washington. He wondered what sort of mood Grace would be in…

## CHAPTER EIGHT

Grace was cool when Walton rang. Why hadn't he called her before? In the old days, she reminded him, he'd always phoned her on arrival at the airport.

He explained that the plane had been delayed and he was bushed. Couldn't she understand that? She could, but when she'd called his flat the phone was engaged. He could at least have called her first.

Walton was cornered. He'd called Lestari first, knowing she'd be anxious about his return. Grace should have known about the flight delay.

He countered. Her phone too had been engaged. Grace told him Powell had called, asking her out for dinner. She'd accepted.

"But you knew I was coming back tonight -- how come you're seeing that guy?" Walton was angry now. Grace too was annoyed. "He asked me out -- you didn't. That's why. Anyway, I thought you weren't going to call."

"Can you cancel it?" asked Walton.

"No, I can't now."

"Why not?" asked Walton, a demand in his voice.

"Because it's too late -- and anyway it's about time you realised you can't just walk in and out of my life."

"But, honey I'm not back for long. I've missed you, you know."

He tried again, more light-heartedly "I'm not designed to sleep on my own."

She didn't see the joke. "Trust you to mention that first ....an easy lay when you get back into town. Is that why you called?"

Walton denied it. "Honey you know better than that. Look I'm probably a bit jealous. It shows I love you -- I've never been like this before."

Grace was unrelenting. "You mean I've always been at your beck and call before. Well you're not taking me for granted. I've said I'll see Bob and I'm sticking to that. I'll see you tomorrow in the office."

She softened slightly. "If you'd only called me first it would have been OK, you know that. You get a good night's rest. You'll need it -- there's a lot going on at the moment. I'll see you in the office in the morning."

Walton grunted, annoyed at having to spend the evening on his own. He'd felt like company. He tried another tack. "That guy Powell needs watching you know.

"You be careful what you say to him."

Grace didn't like the inference. "There are other things to talk about rather than PWN and trees you know. At least he's not obsessed with them."

They rang off.

Walton poured himself a drink. Not for the first time he wondered about the aggressiveness of American women. They seemed to be running the place more and more and the men were just allowing it to happen. But, he admitted, he wasn't blameless. As usual he had taken her for granted. But why didn't they understand men couldn't be hearts and flowers all the time -- didn't they realise work was important?

He shrugged it off, bathed and slumped into bed. Maybe, in the end, it was better this way. He could never have managed the romantic bit tonight anyway. But if nothing else he could have done with a massage for his ego.

***

The White House announcement was bald. In a final bid to help scientists the Air Force was to bomb a cleared area of Alaska. It was an experiment to gauge the effects of fire on PWN. The President would be at the air base at noon tomorrow, flying there in Air Force-1.

Powell called the White House, dodged his way past the press aides, and got hold of Joe Simon. One -- could he interview him about the story and two, could he get a seat on Air Force-1?

Joe surprised him. "Bob, I think this time you might get an interview with the President. That suit you?"

Powell joked that he didn't mind at all. He guessed he could squeeze in the President somewhere. Joe surprised him again. "Get 'round here now and I'll see what I can do." Powell yelled the score at the news editor as he ran for the door. Twenty minutes later he was with the President, tape recorder running.

Lean was positive, dynamic and serious. "People may have the idea the situation has calmed down. It hasn't. It's worse. Any day now the serious shortages will start to hit home. I'm advised we have only one chance of maybe finding something to kill the PWN -- and that's subjecting it to fire. You may not know it, but I've studied these things and fire is a natural agent in our forests. Maybe we can give nature a hand."

That was the crucial bit. Lean went on about his own position, his need to give a lead by being at the base, but it was just so much propaganda and Powell knew it. The soundbite would do to head up the evening newscast.

He burst back into the newsroom with only minutes to spare, dived into a booth and barely had time to recover his breath before the bulletin started. He introduced it, stressing it was exclusive and ran the interview in full. He was still live. He put up one finger to the controller indicating he'd finish up with a minute review. He used it to emphasise how the President had told him the situation was far from over yet, how whole teams of scientists were working day and night in a bid to find a solution.

The President hadn't changed his mind since that broadcast a month ago. America was in trouble. The news editor chided him for taking the extra minute, complained that there was too little fact, too much opinion, but let it go. Powell had done well.

Powell though had little time for congratulations. He had some packing to do. Together with the elite of the nation's newsmen he had a seat next day for the trip on Air Force-1 to the bomber base. He was glowing. He was into the big time. And tonight, there was that date with Grace.

***

Air Force-1 was brim full. Lean, casual as ever, strolled across the tarmac, greeting newsmen by their first names. Joe Simon, still clutching a bulging brief case, scuttled aboard later. To anyone with half an eye there was an air of reluctance about him. He'd rather have been back in the office.

Powell made a mental note to tape an interview with him on the way back. The President came back into the body of the aircraft, heading for him.

"Hi Bob, that was quick work yesterday wasn't it?"

"Yes sir, and much appreciated, thank you."

Lean passed by. "We'll talk again Bob."

Powell managed to find Simon, elbowed a press aide out of the way, and sat by him. He teased him about the bulging briefcase. "Full of paper?"

Joe grinned. "It didn't strike me as a time for a holiday, did it you?" Powell nodded. He understood. Simon was obviously opposed to the bombing move. Within minutes of take-off the bar was open.

Newsmen got busy downing bourbon, chatting about almost everything but the story. It was a good break from the office, an easy piece to cover. An extract of the President's speech, a talk to one of the pilots -- pre-briefed by the press aides -- a few sentences of commentary with basic facts and that would be the day's work. No hardship -- and time enough for a few drinks meanwhile.

Powell left Simon. He needed to sleep. It had been a late-night last night. Grace had barely stopped talking. She'd needed a shoulder to cry on and she'd chosen him. Over dinner at first she'd been good company, congratulating him on his scoop, wondering about the ethics of his impromptu piece of comment at the end. This time they'd gone back to her flat. And she hadn't objected.

He'd noticed the razor, a man's, in the bathroom but hadn't said anything. In the living room she'd poured the coffee, was generous with the brandy and then it all started pouring out. With hardly any prompting she bared her soul about her relationship with Walton. There was almost bitterness there, thought Powell.

He'd listened sympathetically, managed to manoeuvre himself next to her on the sofa and let her rest her head on his shoulder. He'd put an arm around her, and she hadn't pulled away. But when he'd bent to kiss her, she murmured 'no' and moved off a fraction. She'd looked up at him, put a finger to her lips and transferred it to his. "I couldn't Bob. Not yet."

He'd grinned. "I just thought you should know I'm ready and waiting."

She smiled "I know."

***

Air Force-1 was beginning its descent. There was a bustle aboard now, drinks away, seat belts on. Time to work. The circus was about to go on show.

In his hotel room that evening Walton swirled the ice cubes around in his glass and mockingly toasted the screen. It was a sick joke. He'd been invited but had managed to find an excuse so he wouldn't have to attend. He wasn't going to be paraded out as a down the bill caricature of a scientist. Not even for the President. If Joe had asked him, then maybe, but from his talks with him he knew what the day would be like.

Idly he watched the CNN report. Military bands, inspection of black hulled bombers, close ups of the bombs, an illustrated hoarding of PWN itself.

Cynically Walton remembered an old Forest Service slogan from way back "Woodsy says -- trees are for our use, not abuse." He wondered if Lean had ever looked up that one.

There was even live action of the bombers over the target area, in slow motion too. Then the flames that followed the explosions, the slow rising to heaven, pall after pall of smoke. A funeral pall?

Over all the pictures, patriotic music, cutaways of the President shaking hands with the crew then Air Force One taking off into the skies, vanishing once or twice into the smoke as it went.

The host of the evening show came back into vision in the studio.

"So, there he goes folks. A big man with a big job. I kinda guess that takes some guts, to bomb part of your own country. Kinda gets you right here. But if I could tell you a little secret folks. You may have been wondering about the animals in those forests there. Well let me tell you the army's been in there for two weeks now, getting them out of the way. They rounded them up, got 'em into some safe places and let them loose again when it was safe."

All over America, Walton would bet, that same "little secret" was being let out, the predetermined, pre-programmed White House hint to the various TV channels. Walton wondered how nobody had known the Army was in the forests. Didn't anybody notice? Then he remembered the restricted zone. Maybe they were restricting information too. Odd, he'd thought there was always freedom of information. But maybe that was in normal times. Maybe Grace would know. He would bring the subject up when they went for dinner…

She didn't know. And she resented being asked. She told Walton "I don't ask Bob about his work and he doesn't ask me about mine. And, by the way, just for your information we're not having an affair, but that's my decision, not his."

"So, he is after you then. I thought he was." Walton was accusing.

"If you mean by 'after me' that he cares for me I guess you're right. But I can't help that can I?"

"I think it's gone far enough. Either you're mine or you're not."

Grace seethed. "How dare you. Strikes me I'm yours as you put it when it suits you."

Walton calmed down. This was getting too heavy. Next thing he knew she'd be talking about marriage again.

He reiterated again. "Darling you know how it is. I told you before I could never make any promises. I couldn't make them then let you down. All I know is that I've never made as much effort with anyone as I have with you. If I didn't care I couldn't be bothered."

Grace eased up a bit. "Dick -- you're the only man I want. You know I love you. All I'm asking is that you treat me properly. Just don't take me for granted."

Walton felt he was doing better now. They had dinner and the rest of the evening was more like old times. He'd assumed she'd stay the night and was surprised to find she'd not brought her overnight bag.

"You are staying aren't you?" he asked.

"Do you want me to?"

"Honey -- you know I do."

"Then why didn't you ask me?"

"You just presumed I'd be sleeping with you" she pointed out. Walton was at the point of getting annoyed about it. Was it really worth all this hassle?

She'd been half teasing -- but only half. When she'd fetched her bag, she re-joined him on the Chesterfield. He held her closely for a long time, waiting until she snuggled into him. He kissed her long and hard and only then slid his hand to unbutton her blouse. Gently he caressed her. Soon she was all his. Then it really was like old times again. Or was it? As Walton turned over and went to sleep Grace lay awake, an unease in her mind. What was it nagging away at her?

Emotionally, she knew, she was his. Even if she wouldn't admit it to anyone else at least she could be that honest with herself. But, in these clearer moments in the night, her mind still had its doubts. Either way, it told her, whether she stayed or went there would be pain.

***

It was 8.30 next morning when Lean bustled triumphantly into Joe's office, waving the front page of the Post. "How about that? The whole front page" he chortled.

Joe forced a smile. "Yeah they all seemed to like it. TV too, even radio."

Lean bubbled on "It's kinda lifted the people too you know. They know we're doing something for them."

Joe had to admit he was right. The comment on the morning phone-ins had been almost all complimentary. Maybe Lean did have his finger on the national pulse.

"We got any results yet? You heard from Walton?" asked Lean.

"John, we had the results before we started. They'd already tried fire, higher and lower temperatures, the whole bit. It was never going to work and we both knew it. Walton's heading back for Indonesia. He can get on with it better over there. I guess he'll be there sometime soon." He paused, to emphasise his point. "Don't forget people have been trying fifty years or so for an antidote so don't expect results overnight. It doesn't happen that way. Don't kid yourself there's a miracle cure around the corner because there isn't, and we have to accept that."

Lean grunted bad temperedly. "I guess you're right. You must be. But we could do with some good news follow up. You got any?"

Joe shook his head. "It gets worse all the time. We're going to have to re-cycle a load more paper. And we can't afford to have two newspapers in the same cities any more. It'll have to be one paper per town. No way out of it. There's that and a whole lot more we have to do and fast."

"No way out?" asked Lean.

Joe shook his head. "No way out." Painfully Joe went through the list of what had to be done. First there was the compulsory merging of newspapers, then the setting up of emergency re-cycling dumps, voluntary at first, compulsory if that didn't work. All shops would be visited by state officials and any non-hygienic wrapping would be confiscated. And no compensation. Trivia such as record covers, candy bar wrappings would all go. All sales of wood had to cease. Receipts would be outlawed; cheque books would not be renewed. Banks would no longer send out statements – it could all be done online. All junk mail, all mail order would be banned. The postal service would have the authority to open, inspect and, if necessary, re-cycle any communication they thought unnecessary.

Lean gasped. "You can't. Half of it makes it sound like a police state and as sure as hell it's interference with privacy and personal liberty."

"Find us more paper, new trees, whatever and we don't have to bother" said Joe.

"There'll be uproar" said Lean.

"I guess so" said Joe.

Joe looked up. "There's more."

Lean gazed at the wall. "OK give it to me. I might as well know the worst."

"We're going to need state funds to set up stocks of pharmaceuticals. Mostly they come from abroad and, incidentally, 40% of them derive from plants in the first place. While we still have the money, we have to lay in supplies. I'm hoping the World Health Organisation will help us. We've helped them plenty."

"What's wrong with the doctors buying them and selling them like always?" asked Lean aggressively.

Joe explained "There's so many unemployed now and there'll be more without health insurance now that people just don't have the money any more. And with the dollar on the floor the price has gone through the roof."

"What sort of things are we talking about?" asked Lean.

"The birth pill" said Joe. "The raw material for that comes from one plant and it only grows in Mexico. I guess we have to get in supplies of that don't we?"

Lean smiled "I guess so. Even I'd vote for that."

"And there's more" said Joe. "We're going to have to put the Army, and maybe some of the unemployed, to work de-silting the dams. The forest service says that erosion is setting in fast, there's already heavily increased silt flow in the rivers and in time that could mean dams getting jammed up. OK so it's long term, but we don't know how bad it's going to be so the sooner we start the better. Let's not lose power too."

Lean nodded. "OK so that's something positive we can do. Any more good news?"

Joe gave out with a wan smile. "I don't know if this is good or bad. The timber companies say they could maybe divert some of their stocks abroad to their home base. But they'll want paying. Now there's only one way we can raise that sort of money."

"How's that?" said Lean.

"By cutting imports of something else."

"Like what?"

"Oil"

"Joe, you're kidding. You mean ration gas too? You think things aren't bad enough?"

Joe had it worked out. "It doesn't have to be rationing as such. We just import ten per cent less. Consumption has dropped over the past few months anyway, so it shouldn't be too much of a hardship."

"But gas rationing Joe -- you remember the fuss in '79?"

Lean looked at Joe's figures. "What you're saying is that it's a choice between wood and oil? Which is more important?"

Joe thought. "If you'd asked me three months ago I'd have said oil. Now I'm not so sure anymore. There's one other factor in this too."

Lean listened.

"It's plastic. Now if I've got it right we have a couple of choices about how we make that. You can do it with oil or wood -- either way. But if you haven't got either then you don't have plastic. Until now, with oil prices so high there's been a switch to using wood by-products for it. Now they reckon we might have to switch back to oil -- and with the dollar as it is that's expensive."

"Unless we cut down on plastic production too huh? Is that what you're saying Joe?"

"Yeah. I reckon those are the options you've got."

Lean stroked his jaw. "But, anyway we look at it, people are going to get hurt; they're going to have a hard time." Joe was about to comment but Lean silenced him.

"Joe, these supplies the timber companies are talking about. We have to pay for these? And the companies fix the price? They'll bleed us dry."

Joe agreed silently, spreading his hands wide. "They could always sell elsewhere."

Lean was thoughtful now. "Joe. I think that's not in the true spirit of these times. It's blackmail and it's unpatriotic. Now these companies own those stocks, for all we know we've helped them with tax credits so they can expand abroad. They're just kicking us in the teeth."

Joe nodded. "It's called free enterprise. We believe in it you remember?"

Lean drew a deep breath. "But there has to be a line between profits and exploitation. I've made up my mind. This is what I'm going to do. I'm going to tell those greedy sons of bitches they have a choice. Either we get that timber for the transport costs only -- and we'll compensate them when we have the money -- or they'll find their taxes are going to go so high they'll be bust by the end of the year."

Joe whistled. "You know how much clout these big companies have? You know how many friends they have?"

Lean put up his hand. "I can get people with me on this one if we move quickly enough. And I do have another option, another stick behind my back."

Joe was intrigued at the quickness of the man's thoughts. Lean continued "In a time of national emergency I can just damn well commandeer these companies if I feel like it. And if I have to I will. OK, so I can write in some compensation terms for later on, but if I have to I'll wind up those companies, take them over, call it nationalisation, and they can scream all they like."

"That's kinda tough John."

Lean grinned suddenly. "Yes it is -- and that's the way we're going to play it from now on. I might even go on TV again. Just get me all those orders drafted. I'll go talk to people on the hill. They have to be told what we're facing. They have to know who's running this place. OK?"

***

Within minutes of leaving the aircraft Walton was sweating. He looked anxiously for Lestari. He saw Nani first, running towards him, clutching a doll. He picked her up, cuddled her hard and laughed as she smothered him with kisses.

"I'm glad you're back" she whispered. He had that lump in his throat again. He couldn't speak. What was it about this kid? He sniffed, looked up and around for Lestari.

She was about ten yards away, watching. A quiet smile on her face, a touch wistful maybe, but warm too. Dick saw her, smiled broadly through a slight mist and put Nani down. He faked a sneeze to give himself a chance to pull his forearm across his eyes.

Lestari wasn't fooled. "I think Nani can embarrass people sometimes. She is so full of love." Walton ached to take Lestari in his arms. There was no chance. Nani was hung on tightly to his one hand, and his case still in the other.

Lestari half-turned, half-lifted her head and the chauffeur relieved him of the case. She quickly squeezed his free hand, looking him full in the eyes. "It has seemed a long time."

Walton's stomach was tightening. He just couldn't think of words that would fit. He just nodded, slowly closed and opened his eyes, took a deep breath and muttered "Too long."

Nani was chattering away, half-skipping with excitement. She let go of his hand and ran to the car. Lestari walked close to his side, brushing against him. He could hardly keep his hands off her. He smiled ahead at Nani and without turning his head told Lestari.

"You'll forgive my saying so but if I don't make love to you within about the next five minutes I think I shall explode."

She laughed softly. "Mr. Walton sir, that is not the way of the Orient. Here we take our time. I think I can wait at least ten minutes. I have made arrangements."

This time she turned to him. "I have permission from my father to accompany you to your hotel and to brief you on the latest developments."

Walton almost laughed out loud with joy. "Do you think they'd find it acceptable if I jumped in the air and cheered?"

Lestari giggled. "Only if I did it also."

"But what about Nani" he asked suddenly. Lestari answered "Usually she would now be having her afternoon sleep. Because she was so excited I allowed her to come with me to the airport only. I told her it was that or nothing."

Walton got in the car, Nani wriggled on to his lap and they hugged each other. "Did you bring me a present?" she asked impulsively.

Lestari scolded her. Walton asked dourly "Have you been good?" Nani nodded, her thumb in her mouth.

"In that case" Walton teased "I might have something. If you are very good and have your sleep I shall give it to you this evening. OK?"

She nodded, the car dropped Walton and Lestari at the town centre hotel. He checked in, the clerk queried if the booking was for one or two and Walton emphasised that he only was staying but that he and Lestari would be working during the afternoon.

The room of the suite closed after the bell boy. Walton drew Lestari to him and led her to the sofa. "How long can you stay?"

She smiled. "I think it will be long enough. You must be tired though?"

Walton joked "If you mean do I want to go to bed you're right, but not alone."

Lestari shook her head, Walton looked alarmed. She smiled. "I think you should have a bath. Shall I run it for you?" Walton looked a bit put out. And urgent need was surging through him. He had never seen her look more beautiful. He shrugged his shoulders. "OK, that's fine."

He sank back wearily into the chair, hearing the bath water run. It seemed to take forever. What was she doing? He stood up as she reached the door. Her hair, black against the deep brown of her back, swayed as she turned away from his lunging hand.

"Dick" she laughed "you must not be so impatient. Get into your bath."

He stripped off in the bathroom and stepped into the bath, somewhat embarrassed by the size of his erection.

Lestari came in, knelt by the side of the bath and quietly collected his clothes. Walton leaned back thinking about the virtues of Oriental women. He couldn't see Grace doing that.

Lestari re-appeared in the doorway, picked up the soap and beckoned him to sit up. Gently she soaped his back, softly massaging it with her hands. "Now lie back" she said. Tantalisingly she soaped him all over, caressing him, slowly stroking him everywhere. She stood up and, as he watched, the towel slid over her naked breasts to the floor. Silently she stepped into the bath with him. "Now you wash me."

Walton grabbed the soap with gusto, lathered his hands and lingered them over her body. She was so firm. Her breath came more quickly now. She slid her hands along his legs to caress him again. Walton stood up, almost dragging her out with him. Still dripping, they stood and embraced. He took her hand and led her to the bed. This had to be his woman. She lay back on the bed, stretched sublimely, an erotic vision of a wanting woman. At that moment he was as sure of it as he'd ever been about anything.

"Kiss me first" she said. "Everywhere." He started from her ankles. As he worked his way up she moaned as his tongue touched her clitoris. "Oh Dick -- now." And in a second he was in her, sliding through her wetness. To oblivion. This time it was truly lovemaking.

He awoke much later and found her smiling gently into his face.

"Dick, I have to go, but I couldn't until you woke up. You looked so peaceful. My father is expecting you at eight."

Walton half laughed. "And what sort of briefing shall I say you gave me?"

Lestari giggled. "I think perhaps we could say it was a briefing of mutual satisfaction."

Walton nodded "I guess you could put it that way."

He felt at peace. All the aggro of America, all the pressure had gone now. It was all behind him. It was a different world. He pulled up his thoughts short. Oh God it wasn't. It wasn't at all. In all his eagerness to make love to Lestari he'd never once asked how the country was coping.

His worry showed. Lestari's brow creased for a second too. He pulled himself up on his elbow. "How are you coping? I'm sorry I didn't ask. It was just that....."

Lestari shushed him. "I'll leave you some notes. It is bad, very bad." Then, briskly she left.

Walton slumped back and lit a cigarette. If only it could have been different. He dozed, only to be woken by the phone. It was Grace. "Hey warrior, I thought you were going to call me when you arrived there. That was hours ago."

Walton apologised. "I had to go straight to a briefing. They work fairly hard here too you know. They don't waste any time." He quickly remembered to ask how she was, had anything happened in the last few hours.

She was guarded. "Dick, you know the way the phone can be here right now, but yes, things are moving fast. I guess you can get the gist of it on Voice of America, but there's one thing you should know because it will affect the people there. There's talk of maybe the White House taking over the timber companies unless they get some wood over her fast from their bases abroad. That's all I can say really. Try to get the rest on the radio, it'll all be in Bob Powell's piece."

Walton felt a pang he couldn't recognise. "You still seeing that guy? I thought it was just business."

She was offended. "I still see him yes, and you're right it is just business. But at least he calls me when he's asked to." She's back into the old speech, thought Walton. Would she never give up? Lestari wouldn't lecture him this way."

He interrupted her. "Listen I have to go -- got a date with the Vice President." She said nothing, just asked him to call when he had the time. Sarcasm never did suit women, thought Walton.

He checked his watch, went to his case and pulled out his battered old radio. He found the Voice of America frequency and took the radio into the bathroom with him. In ten minutes, he was listening to Powell on the news bulletin. And now he knew what Grace meant. A whole string of measures had poured out from the White House. There was the clear threat to nationalise the timber companies, a panicky sounding item about pharmaceuticals and a piece about inspectors retrieving superfluous gift wrapping. It sounded more like a police state than America. He wondered if Lestari's father had heard it yet. At least now he couldn't think America wasn't taking things seriously. But wasn't it time that they did? And then again couldn't they have started just some of it years before? He noticed too one omission from the bulletin. No mention of the bombing raid on Alaska. No results. Nothing. But perhaps it was just as well.

He stepped out of the bath and checked the air conditioning. Maybe it was his imagination, but it seemed a lot hotter here than on his previous visits. He grabbed himself a drink, appreciating the cool of the ice on his palms.

There had been the mention too, he now remembered, of the Army working on desilting dams. It was ahead of time, but it would be an effort well spent. He'd seen it happen too often in other countries where trees had gone, and deserts had spread. It all happened so quickly. And this, he thought, was the most powerful nation on earth. How could a relatively poor country like Indonesia hope to manage a crisis of such proportions?

It made him anxious about his meeting with Lestari's father. Walton wondered how he should play it. Maybe Lestari would help him there. Quickly he dressed to be ready for her. He dived into his case, unearthed the doll he'd brought for Nani and poured another drink.

She was prompt, but the chauffeur was with her. Their greeting was formal. In the car heading for the house he fired a series of questions at her. All the answers were bad news. They could hardly have been worse.

Like an old country doctor, he recognised all the symptoms. The country had been served with a death warrant.

## CHAPTER NINE

The old man was solemn, his greeting polite but formal. He'd heard the news, but had some more for Dick, something that had not been on the bulletin. America was informally to cut its oil imports by ten per cent.

Walton was surprised and, with a touch of irony, the old man explained that without such a cut, production of plastic would have to be restricted. Now Walton understood the irony.

In Indonesia the people in the jungle would be hard put to find wood to cook food, in America they bothered about plastic sacks into which they put their piles of leftovers.

"As you know" the old man said "Oil is our main export and we would hope that America would still buy the same amount from us. Could that be put to your government?"

Walton wondered for a second if it was his job, he wasn't an ambassador, but he nodded anyway. Lestari's father was also curious about the bombing raid on Alaska. Was there really hope? Walton shook his head and explained that behind it was the President's idea that the people were in need of leadership, that they somehow had to be impressed with the size of the crisis. It was hard to explain but.....

The old man lifted his hand. "I understand. As the President said you are a young nation and, like a child which has always had its toys, you cannot understand that in times of poverty there is no money for playthings. A child can only understand when the toys are taken away."

Wistfully, he added "We have no such problem here. Here poverty is understood."

"Aren't there other differences too?" asked Walton.

The old man raised his eyebrows in a question.

Walton went on "The more I see of America, the more I see greed, selfishness and an uncaring waste. It might sound disloyal, maybe it is.

But here at least there is some caring for others. There is sharing amongst the family."

The old man agreed, but added "You must not blame yourself, or even your fellow countrymen. When there has always been plenty there has not been a need to share. We are used to that need. It is just that we are older, that is all. Your country will learn."

His face took a more serious countenance. "I also understand that your President is demanding the American timber companies send their stocks home and, if they don't, he may take them under his control?"

"I've only heard the bulletin, but that could be right."

The old man sounded sad. "My President cannot allow that to happen. We too have to be a little selfish. Remember, a lot of our people need wood for heating what little food they have.

We cannot take the trees from the jungle to the docks, past the faces of starving children and hope they'll understand the trees belong to Americans and not them. They understand much, our people, but that is something they couldn't comprehend. You see what I mean?"

Walton understood. He resented slightly being lectured about the ills of America and its society since he'd spent his life urging conservation. But he had to sit and take it.

Gently the old man continued "Mr Walton. I don't know how much you know about our country, perhaps more than most of your fellow countrymen. But don't get the impression that because we have oil we have wealth. Our original fields are drying up and the new ones are difficult and expensive to exploit. We have no money or expertise -- for that we have to rely on foreigners. That means that not all the income is ours. And yes, we are the world's leading exporter of tropical hardwood, but the real money in that is to send it abroad as processed articles, not just as raw logs. We have tried to do this, but that too takes money and expertise. So once again we have to rely on outsiders, the Japans and the Americas. And they are in the business of profit. And while our land is fertile do you realise how much rice we have to import to feed our people? It's about a third of all the rice in the world and, again, that is not cheap. Do you know the average income of our people? 300 dollars a year. Most Americans earn that in a few days, a few hours maybe. We are ten thousand miles from New York, but it may as well be a different planet. Your companies scoff at us, complain about our civil servants taking punglis -- you call them bribes -- but how could they take them if they were not offered?"

He was bitter now. But Walton could understand. He remembered the headline from an American business magazine feature on Indonesia "Wealth waiting to be exploited."

Lestari, sitting quietly on the floor, gently interrupted her father. "I think Mr. Walton knows how things are. I feel he is here to help He has come a long way to do that."

Walton smiled at her gratefully. It had been an earful. He turned to the old man again. "Sir. You are right that most of my people do not know these things. Many of them would not know where Indonesia is on the map. But if they did know I'm sure they would care."

"How sure are you of that? And what would they do about it?"

Walton accepted the point. "I know what you mean, but I'm sure that in more normal times the government would at least ensure aid in some form or another. We have done so in the past."

"And in return" said the old man "there has been a political advantage in doing so has there not?" Walton had to agree.

Now the old man looked wistful again. "You know of course that there will be those who may take advantage of your plight."

Walton looked puzzled. The old man explained "The Chinese community here is strong and always seeks to extend its influence. Japan too, which has been our best customer for wood, now seeks a contract for all our rubber -- and at good prices. And the Russians too, like Americana in the past, have always known that timely foreign aid is always appreciated by a poor country."

Walton felt he was getting out of his depth. This really was stuff for ambassadors. He nodded vaguely. "I think I understand what you're saying sir -- that this crisis obviously hasn't endeared us to you. That we may have lost a friend?"

He had to explain "I am only a scientist sir.....here to try to make amends for an appalling mistake, albeit an accidental one. Should you not say this to our ambassador?"

The old man sighed slightly and explained, as if to a child.

"Mr. Walton. If I, or my President, talk to your ambassador it is country talking officially to country. Positions are taken and, once taken, can become fixed.

If, however, there are those in America who understand our position in advance then perhaps more can be achieved. I welcome your work here as a scientist. I believe too that you are a friend and as such you should have our trust. As a trusted friend therefore, I am sure you appreciate our position and it would be perhaps unnatural if you were not to mention this to your Mr. Simon. In this way you could also help us. Do you not have an expression about *'a word in the right ear'?"*

Walton had a better idea. "I think the one you need sir is another that came from the English -- it's about *'a stitch in time'*."

The old man relaxed now. "Mr. Walton I think you have earned a drink. And I also think I shall be in grave trouble if I don't let in my granddaughter to see you. It seems you have some business with her as well."

Walton laughed. "Of course. How could we have ignored something so important for so long?"

Lestari fetched Nani. The little girl rushed at Walton, almost bowling him over, asking him if he'd brought her present. Walton feigned ignorance, teasing her, asking if she'd had her sleep.

Lestari, unable to resist the private joke, said "Yes -- and she helped her mother. She had her sleep."

Walton handed her the doll and Nani cuddled it, hugged Walton again and kissed him on the cheek. Lestari, fixing the drinks, looked over them to her father. His eye caught hers, she smiled, and he nodded with approval.

Lestari relaxed. During the lecture her father had been so stern, she wondered for a minute if some of the remarks about exploitation had not in some way been a subtle warning for Walton.

She wondered how much she meant to Walton, there was a woman's instinct at work in the back of her mind, but she'd already dismissed it. It was fine to have a man again. And it was good for Nani too.

The old man, still enjoying the sight of Nani romping now on the floor with Walton, coughed slightly and Walton looked up.

"I think Mr. Walton perhaps you could be more comfortable -- if you preferred it of course -- staying here with us. We have a guest cottage on the grounds. It is simple, but it is comfortable, and it would be at your disposal."

Without even a hint of a twinkle in his eye he added "You would have the privacy perhaps you need and that would not be available to you in the room of a public hotel. It would also have the advantage of security should any of your work require a degree of secrecy."

Walton loved the idea. The work part made all sorts of sense and, though there'd not been the slightest hint of it, he was sure there was tacit approval of his relationship with Lestari. Graciously, he accepted the offer; the old man said Lestari would make all the arrangements and Nani jumped with glee.

The old man teased his granddaughter for forgetting him and, solemnly, she offered her doll for his inspection. She sat on his knee as he explained that Walton was a guest and part of such hospitality was the courtesy of privacy. Visits to the cottage would be by invitation only. She reluctantly agreed, kissed the old man and Walton then went off with Lestari to be put to bed.

Alone now the two men said little. They were from alien cultures and different generations, yet there was much they shared, that they understood, grieved about and hoped for.

As Lestari returned the old man excused himself. Walton thanked him for the cottage -- and meant it. At least now he and Lestari had the privacy they sought. And that night they made the most of it.

***

This time it was Powell who suggested he should go to Detroit. It meant being away from Grace, but, he had to face it, she was still too loyal to Walton. But, beyond that, Powell felt he had to get away from the remoteness and unreality of Washington. It was all talk; the crisis had an air of aesthetic unreality. He needed to see what was happening on the street.

The editor wasn't too happy about the idea. Just lately Powell had established himself both on the air and in the White House. People were paying attention to what he said, the guy was even developing a following and there wasn't another reporter with contacts like his on the Hill. But, he indulged him anyway. He didn't want him going stale.

Next morning Powell's plane began its descent, circling down over Lake St. Clair. He stared down intently. Where each river spilled into the lake there was now more than ever a stain of brown; the silt was increasingly being washed away from the hills. And every lake was the same. God knows how many extra tons a day of fertile soil was being wasted. And there was something else about the lake today. It was crowded with boats -- dozens more than usual. Despite the chill of the spring there were small armadas, cruising all around, trailing fishing lines behind them.

The man in the next seat knew what it was. "If you ain't got work, you fish and fill up the freezer against hard times."

Powell noted the observation; it would make a comment on the times. It was the sort of human touch that would make people relate to the crisis.

"They hunting too?" he asked his neighbour.

"Sure" said the man "but it's different now. They can't use scatter guns as much anymore -- they don't have the cardboard for the cartridges and the plastic ones are getting more expensive all the time. But straight bullets, yes they still hunt with those, when the rangers will let them into the woods."

Powell was curious. Wearily the man explained "Say man -- where've you been? Haven't you heard the state is trying to stop people poaching trees? Don't you know there's whole gangs at it now? You been on the moon or somewhere?"

"No" said Powell "just Washington."

"I guess that explains it" said the man, bitterly. "They never did know much down there." Powell couldn't help but agree.

The plane landed, he hired his car. At least there was no paperwork now, not even a receipt for his expenses. The girl warned him in a parrot-like way about the dangers now of walking alone downtown at night. She told every customer that.

"But wasn't it always that way here?" asked Powell.

"Yeah" she said "but there ain't so many folks got jobs now. And the welfare doesn't last forever you know."

***

Powell drove off, heading East, using both his wipers and heater. It was a typical Michigan spring day, wet and chilly. But, somehow, the air along the freeway was clearer now. There was less smoke staining the air above the factories along the freeway fringe. Maybe it was for the good, but it meant a lot of jobs had gone in the instant recession. What a choice. Pollution or jobs. Slowly now, he cruised along in the slow lane. There was another difference too and he couldn't quite twig what it was. It wasn't just that the air was cleaner. He was seeing more for other reasons too. What was it?

A boarding caught his eye -- and now he realised what it was. It was the only one that had survived.

All the giant slogans that proclaimed the commercial offerings of the city, all the testimony to the American way of life had been swept away. The place had to be the better for it. But it took some getting used to.

He swung off the freeway and into the downtown area. He glanced left at the Tigers' Ground, looking almost naked now without its posters for forthcoming games. The whole look of the place had changed.

At least the graffiti was still there, mostly anti-Lean. "Be Mean with Lean" still took the top marks, echoing the cry of the UAW from the other side of town. They'd argued for priority supplies for the auto industry. Lean had said 'no'.

He flicked on the radio, a professional ear at work now. It was a commercial. He could hardly believe it. It was for Joe Muer's, one of the city's most famous restaurants. After what seemed like centuries of only allowing its customers to pay with cash, the place was now actually taking credit cards. Things had to be bad. But, he wondered, how long would the plastic for the cards survive.

A time check reminded him he had no need to book into the hotel yet. He'd wheel up Woodward. Just beyond the junction with State Fair he glanced, surprised, to his left. It was busy with cars in the approach roads. He slowed and made a left. He got out and gazed at the cars. They were lining up to deliver paper garbage to the city re-cycling dump. This was new. He collected his recorder and went along the line asking people about their motives, what made them want to help in the effort.

"It's like this" said one. "The President says we gotta help ourselves and we're doing just that. It don't cost us nothing."

A second agreed. "What have we got to lose? At least this way we have something to do. We sure as hell don't have jobs anymore."

The man's wife took up the point. "And if we don't bring it here, then I tell you boy, you'll just have somebody breaking in to take it so they can sell it to a pirate dump."

"A pirate dump?" asked Powell.

The man looked puzzled. “You heard of those surely. They’re all over man. Take anything there and they don’t ask no questions. And some of them pay real good money for a door or a shed. It’s a whole new industry.”

“Where would I find one?” asked Powell.

“You’ll find them soon enough” said the man.

“So why don’t you go to them?” asked Powell.

The man turned on him almost angrily. “Just because I fought for this country once; and when the President says he wants something done then this guy’s going to give him a hand. That’s why. Don’t they teach these things where you come from?”

Powell thanked him and went back to the car. Maybe there was still some patriotism left. It came as a shock to his cynical soul.

He re-joined Woodward, heading out towards Six Mile, the start of two miles of Vice quarter. He stopped halfway along but had left his recorder in the car. as it made him noticeable. The policemen across the road glared to see if he’d parked in a legal zone. Powell walked along the road, registering how many people now seemed to be crowding the sidewalk. There was a line outside the soup kitchen, but since when was that unusual? It was almost a hallmark of the bad times the city had known.

He passed a hooker who had on her professional friendly smile.

“You wanna go for a while?”

Powell shook his head. “Sorry ma’am, not today.”

“Oh, come on boy, you look real smart. You ain’t short of cash are you?”

Powell shook his head silently and eased past her. Her face was slashed with vivid red, widening her already full lips. They stretched into a knowing grin. “I take credit cards you know.”

He turned back to his car and drove off quickly. He felt unclean somehow. Ahead of him the flashing lights of a police car halted his progress. Three officers had their guns out, stood behind the doors of their car. Ahead of them an old lorry loaded with doors and old packing cases. Powell pulled over, felt for his press card in case it was needed and walked with his recorder moving to behind the police car.

One officer turned, both hands on his gun. “Get outta there. Get lost.”

“Press” yelled Powell.

"This ain't no story boy" shouted the officer, wheeling around to face the lorry.

Powell went back to his car and waited until the officer had checked the lorry. One of the men was escorted to the car and put with the officer in the back. The car drove off, shepherding the lorry in front, now with one of the police officers, gun at the ready in the passenger seat.

He followed it to the precinct station. The officers went inside. In the yard were four such trucks, each with its booty of door and cases, with here and there a freshly cut tree trunk.

One of the officers re-appeared and came over to him. "What's your problem?" he asked Powell. Powell flashed his card and asked him how common such incidents were. "Now come on man, you must know. Happens all the time. Wood's the new gold. Hadn't you heard?"

The new gold... Powell noted the phrase, thanked the officer and drove downtown to book into the Radisson Cadillac as usual. It was a different world out there, so many extremes of behaviour along just one road.

In what other road would you get patriotism, credit card hookers and traders in the new gold. The new gold ... he liked that phrase. He guessed the editor might too.

He called him, filling him in on what he'd found out so far. The editor grunted. "Sounds a bit far-fetched, but if you're sure about it, then OK."

Powell put down the phone, waited for it to ring, and connected this time to the studio hook up. He read over his piece, feeling like a war correspondent checking in from the front. He guessed there'd be more like this.

He called Grace, asking her to listen to it on the evening newscast. He thought she'd like it.

Jesus -- she sounded so warm. And even pleased to hear from him too. Was there any news of Dick? As usual she chided him for asking that same old question but said 'no'. She'd tried calling his normal hotel, but, they said, he'd moved out. No, they hadn't known where. Powell tried hard to sound sympathetic, but it was a feeble effort, so much so that Grace had to laugh.

"I guess what I really mean" he said, "is I hope he's found some dusky maiden and gone off to Bali with her."

Grace's laugh this time sounded a bit more forced. Powell caught it – he could tell she was really worried about that. Mind you, thought Powell, they did have some beautiful women out there. You could hardly blame a guy.

Grace too was beginning to know him. "You sound sort of envious Bob. Are you?"

He teased her, putting on an act of unrequited love." Honey, you know me -- I'm just waiting for the big moment. And it'll come you wait and see."

"I guess it might at that" said Grace almost absent-mindedly. Powell's eyebrows went up all of their own. "You mean I stand a chance in a million?"

It was Grace's turn now to joke. "Heavens no -- I'd have to work my way through a whole heap of others first."

"I'll stand in line" said Powell. And he meant it. "Now you just go and listen to that bulletin and realise what class you're mixing with nowadays." She just caught the start of it.

The reaction to Lean's tough new measures had been surprising and varied. From commerce and big business there was anger. The freedom groups were suing everybody in sight over what they said were threats to individual liberty. And the rumour about the cut in oil imports was causing some panic too. By and large the churches seemed to be backing Lean. And a lot more people had been there of late.

Then came Bob's piece. It was good -- the story of new gold, how some people sought it and others put their country first. A time for each person to make their own decision.

No doubt about it, thought Grace, the man had a feel for the human story. Maybe it came from his own warmth. And he was that. It was perhaps that trait which most appealed to her. He'd been considerate -- and patient too. Now Dick -- he'd have just pushed off and found someone else.

The phone went. It was Joe Simon. He'd thought Powell's piece showed signs for hope, had she had the same impression? He repeated his invitation for her to drop by and see Ruth and the boys. They worried about her on her own, particularly with Dick away.

Had she heard from him? She told him about the call to the hotel. "Didn't you know -- he's moved to the Vice President's place. It's more secure there and he has more privacy. And he was on to me this afternoon -- didn't he talk to you too?"

She hated admitting that he hadn't. Had it been important?

"He should have been in the diplomatic service Grace. The Vice President tipped him off about what the Russians, the Chinese and the Japanese were up too. And guess what -- they aren't going to release any of their timber to our companies. It just doesn't get any better does it? We could really have done with that stuff. Say, I'll get you the number, I bet he was real sorry he couldn't talk to you."

She took the number and thanked Joe. She'd take a rain check on seeing Ruth tonight. But right now, she'd like a night on her own.

"Don't tell me you're washing your hair" joked Joe.

"Something like that" said Grace, "Just getting things washed up generally."

## CHAPTER TEN

Powell swung the car onto the forecourt of the filling station and winced at the price of gas -- that's what the low dollar was doing to them. Who needed rationing when it cost that much?

All the stuff he'd heard about the plummeting value of the currency really hit home here. That fall in the dollar hadn't pleased the Arabs either. With America cutting down, their revenue too had fallen and, increasingly now, they were pulling their money out of the country.

The tank filled, he offered his credit card. The attendant shook his head. "We gave 'em up sir. Sorry."

Powell cursed. It was the second time today. "What's the problem?" he asked. The man explained, weary of what had become a set speech now. "For a start we had to call the company each time and tell them who the customer was and how much the charge was. Then they'd have to look it up to make sure the customer still had some credit left. It cost a fortune on phone calls and then half the time they'd get the payment mixed up or the customer would get charged the wrong amount. It just wasn't worth the hassle any more. So now it's cash. Didn't you see the notice?"

Powell followed his finger to the chalked sign, just two words "Cash only" It said it all. He pulled out his wallet, extracted the bills and paid the man. Again, there was no receipt for his expenses claim. The whole country was running on trust.

As he cruised out along Telegraph towards Pontiac he noted the incident in his mind. That had been just one small transaction. God knows how the banks and Wall Street were managing.

He'd heard on the news reports that almost everything now was operating by word of mouth. What was really fouling up the system was that the big corporations, having switched over decades ago to computers, were now in trouble with their accounting.

Computers needed paper and there just wasn't any around -- at least not in enough quantity. Even the Treasury was finding problems in getting paper for dollar bills. When that ran out he dreaded to think what would happen.

His cheque book had long since been exhausted -- like most other people's. Since he didn't get bank statements any more it meant like everybody else he had to go online or to the bank where the clerk worked it out manually. Only then did he get any cash and nowadays the banks refused to hand out more than half the money in dollar bills. So far he hadn't started using the fashionable waist bags that some of the younger guys were using; his pockets weren't going to stand up to the weight of the coins anymore. But he knew the banks had bigger problems than his. Interbank payments, credits, money transfers all now had to be delivered by messengers, mostly cycling around the city centres to save on fuel. Message centres had become the vogue since the postal service carried only essential messages now. In almost each shopping mall, commercial concerns had combined to rent a shop, then converted it, staffed it and used it as an informal postal centre.

Some people said it was actually speeding up commerce. Certainly, there was more trust between folk now -- there had to be. Almost overnight e-mails had been given legal authority where in the past it had to be faxes at worst. Maybe there were some benefits in this crisis. Idly he glanced at the newspaper vendor on the corner of the street. Today it was the turn of the Free Press to be the newspaper on sale. Albeit it was only ten pages, but it was better than nothing. Since Lean had ordered that papers couldn't duplicate there had been one of two arrangements. Either the papers had merged for the time being or, like here, the News and the Free Press published on alternate days,

He glanced at the headline. A cynical smile spread over his face. So much for the benefits of the crisis. It was still the crooks who were making the headlines. Downtown, said the paper, a gang had broken into the synagogue, looted it and savaged the old rabbi who'd tried to stop them. He'd rushed to protect his cherished scroll, the vandals had kicked him aside, beat him around the head and broken an arm as he clung to the priceless manuscript. He'd died last night, but his close friends were saying it was a mixture of shame and a broken heart.

There'd been other victims too. An old lady, nearing seventy, had been attacked and beaten while clinging on to her family photo album.

She was still alive, but while she was in hospital the gang had gone back and stripped her home of all its wood. They'd even taken her bed, a family heirloom. Now neighbours were standing guard knowing that once news of the attack spread other crooks would realise the house was empty -- and still had its doors on.

Doors it seemed were a favourite target, particularly for debt collectors and there was an increasing number of those on the streets now. In Oakland, he read, a chapter of the local Hells Angels had set up in business collecting debts. Their methods were simple.

They walked in, demanded the money and if there was no instant payment they gleefully took the debt in kind. They had a going rate for the job. A door was worth twenty dollars, window frames five.

Powell wondered where it all ended up. maybe at those pirate dumps?

It was like reading about a series of local wars. Increasingly now, neighbours were banding together, taking turns to mount guard and patrol with guns around their communities. The crisis was dividing whole blocks of people yet in a way -- if only through necessity -- neighbours were at least showing each other some sort of compassion. And there were always people who had an eye to what they called "profitunity." An advert caught his eye -- one for metal doors. The doors were being made by an enterprising group of workmen made redundant at a factory which used to make typewriters; that was American know-how for you.

He changed channels on the radio, barely listening as he did so. No, he thought idly, I can do without the religious station. But a phrase caught his ear. The announcer was breaking in: "This programme is brought to you by the Earth Mothers of God," What the hell were they? The preacher came on, sounding vaguely like a Detroit version of Bessie Smith.

"And I tell you brothers and sisters, God is purging this land, reaping this retribution on all us poor sinners. Didn't we tell you that judgement would visit us all? Hallelujah and now let's all sing those blues away."

The vibrant tones of a new spiritual throbbed through the car -- new words to an old tune, urging people to work together pray together and so save the country.

The spiritual ended and the lady preacher continued. This was her main message, urging those trapped in the poverty of the cities to move out and till the land. "For there, I tell you brothers and sisters, is our salvation. That's how we made this country great; by growing the corn and each man and his family gathering up the fruits of God's great gifts. So, join us now, the sisters of the soil, the Earth Mothers of God in this great crusade. Just come along and we'll show you how. Hallelujah."

The commercial broke in -- an advert placed by one of the big multi-nationals now heavily into seeds. Powell switched it off, laughing quietly to himself. Was there anything on earth that someone couldn't make money from?

***

He was at Pontiac now, and in the distance he spied the glistening roof of the Silverdrome. Even that had been used last night for a huge religious rally. He drove into its vast empty car park. Any other morning after a huge event like that it would have been almost ankle deep in litter, wrappings from hamburgers, candy wrappings, tissues of all sorts. Now there wasn't a scrap of paper to be seen. It looked so much better.

He cut off down the side roads, cruising past the plush lakeside homes of the middle managers of the motor industry. Where once the pines had protected each home from the gaze of the other, there were now just piles of logs. A state order had decreed that to stop the PWN spreading itself among what healthy trees were left the trees had to be cut down, covered and stored.

The road, never the best of surfaces, was slippery with brown mud spilling over from a nearby stream. The rains of spring, normally filtered and checked by the suburban woods, had nothing to stop them now and down they poured; each stream bearing its cargo of silt. The lakes too, once reflecting the blue of the spring sky, were muddy with the brown stain.

He made back for the freeway, spotting the direction sign to the University of Michigan. The address was a hollow joke -- Evergreen Avenue. He pulled over and made for a fast food bar. He could do with a coffee. He parked, walked in and ordered some. At least now there were real cups, no more of those damn paper and polystyrene things. He'd hated those and as far as he was concerned the quicker they died the better.

A proper plate too, no more of those bendy paper ones that always managed to deposit his hamburger on the floor. The hamburger was greasy and automatically he reached for a tissue, instantly cursing his own stupidity.

The waitress noticed his dilema. She indicated a new device in the corner, a shallow bowl of scented water, rather like the rosewater in Chinese restaurants. It was much more pleasant than tissues which only ended up littering the table anyway. And, he bet, they never get re-cycled either. He dried his hands, paid for the coffee and hamburger. He guessed somewhere there was some guy making a small fortune out of manufacturing those bowls and the rosewater.

In the car the local news was well into its midday bulletin, featuring, what used to be one of Detroit's biggest wing-dings, the annual conference of the Society of Automotive Engineers. There was a gossipy tone to it -- a little anecdote there about the general Motors guy hoping the delegate from Poland would pay the lunch bill. But on a more serious note the trade was struggling. Scores of others had depended on those component firms. It was like the bad old days of the recession in the motor business of the late '70's. And he remembered how they'd compared it to the big one to the '30's.

The old, old saying, *"if Detroit sneezed the rest of America caught a cold",* was being demonstrated all over again.

***

At noon he called the newsroom, from his hotel bedroom, recording his piece for the lunchtime bulletin. He was enjoying the idea of being a pure reporter again. It was just straight observation, not so much of the wheeling and dealing and the kite flying that was the hallmark of his work on Capitol Hill. That seemed light years away right now. Up on the Hill, he mused, they had no idea what was going on in the streets. Remote men in long corridors, hunched over desks that barely saw the light of day. Maybe they should listen to the Earth Mothers of God.

He rang the local radio station to get a cassette of it for the editor. It would at least appeal to his cynical sense of humour. The reporter who took the call was sharp. *'THE Bob Powell from Washington, the guy who did those Voice of America pieces?'*

Powell agreed it was the same guy, flattered they should have heard of him, particularly up here in Detroit.

The reporter had asked him to hang on while he had a word with someone there. After a moment he came back. Could Powell come over and take part in a phone-in that evening, say around five? They could do with a view from Capitol Hill. Powell agreed and smiled to himself. As a guest on a phone-in the boot would be very much on the other foot.

He guessed he'd be in the hot seat. He was right. The calls that came in had some hard questions:

"What was the President doing?"

"Would they get a solution to it soon?"

"How come it had been allowed to happen?"

The questions surprised him, they ranged far beyond the immediate problems that people were enduring. Overwhelmingly there was a feeling that the administration had let them down, that there should have been some warning, why should they be the ones to suffer.

Powell explained that if it was bad here then how did they think the people in Indonesia were coping. The next caller told him "Man I don't give a shit about those folk -- we got our own families to bother about." It was a typical call; most people were just plain worried about how they'd take care of their families and survive. There were the political calls too. Like the man from the UAW.

"Do you think the President realises we just don't have any money anymore to keep up our own welfare plan? Doesn't he know that's running out too? Is he going to extend the state welfare payments?"

Powell explained he was outside the White House, not in it, but it barely mattered. He was from Washington and that was enough. He guessed the President did know these things, but if the caller thought about it, where would the extra welfare money would come from? It could only come from taxes and apart from the sheer difficulty in collecting those you couldn’t support the majority of the people if only the minority were working.

The man from the UAW wasn't mollified by a long shot.

"Why don't he just switch us over from cars to something else we can sell?"

Powell had to confess he didn't know.

The phone-in ended and Powell sank back in his chair.

"I'd sure hate to be a politician" he joked.

The phone-in host handed the show over to his successor for the evening "Golden Oldies" show and they headed for the bar.

***

The man knew his city, knew how it depended on all sorts of raw materials that, until now, nobody had realised came from wood. "Like what?" asked Powell.

"Man, there's an endless list. You know about things like lignin and cellulose." Bob admitted he'd heard of cellulose. His host nodded. "It's about twenty to thirty per cent of wood, this lignin stuff. It gets used in the way they make plastics, polyesters, ion-exchange resins, adhesives, rubber re-inforcers, fertilisers."

Powell whistled slightly, but the man went on. "There's more. It's used as a strengthener for asphalt, dispersants for oil well drilling, tanning agents, emulsion, ceramic processors, vanillin, soil stabilisers -- I tell you it just goes on for ever. And that's just lignin."

Powell was stunned. The man carried on "You want the cellulose list? Rayon, plastic, formaldehyde, hydrolysis to sugar."

Powell interrupted. "All this?"

The man nodded. "And that's just the basic list. Tell me a factory around here that doesn't have some sort of chemical process going that doesn't involve any one or maybe even more than that? I have neighbours and nearly every one of those guys has some sort of chemical or engineering degree. If they haven't been laid off already they're just busting their asses all day and all night trying to find ways around it. But they're the lucky ones. They have jobs. When those plants stop the ordinary Joes are straight to the welfare line. But you know what gets me? One neighbour I know, he got laid off. And he's still working -- using his kids old exercise books -- trying to work out a new way of doing some process without something that comes from lignin."

Powell could hear tomorrow's piece already. Could hear him doing the great in-depth piece on the sheer weight of the crisis. The heart of America’s industrial North East crippled. No wonder exports had dropped; no wonder the balance of payments was out of line.

He finished his drink, collected his cassette of the Earth Mothers of God and got back in his car.

***

At the hotel the maid had his laundry ready and wanted paying. She was big, black, bosomy and looked, thought Powell, like the archetype of an Earth Mother. There was too a pride in her face.

He chatted to her for a while, hoping for a crumb of apparently useless information, something that would bring tomorrow's piece down to earth. He asked how she was coping.

She sighed, a despairing heave that was almost a spiritual of its own. But apart from the rheumatics she was fine. "Couldn't you get them fixed" asked Powell, "don't they have drugs for that now?"

"They sure do boy -- or at least they used to. The doctor man says he can't get them anymore. I gotta stand in a line at the welfare clinic tomorrow and see if they've got some left?"

Powell wished her luck, gave her an over generous tip and wondered why the drugs for rheumatism weren't around. On the off chance he rang his friend at the radio station, saying they could share the story.

His friend laughed quietly. "That's not news here. But I'll tell you what it is. Here's something I never knew, so I looked it up and there's a thing called chlorophyll-caratene that gets made up into a paste and is used for rheumatics. Trouble is it comes somehow from the leaves of some tree somewhere. And there are none of those left. She's just gonna have to pray it gets better, 'cos there's none left in this town."

Powell thanked him, swirled the ice cubes in his glass and wished he had someone to talk to. He stripped off, stepped into the bath and decided he'd call Grace. Even the thought gave him an erection. She wasn't there, he'd try her later.

***

Grace and Ruth were gossiping in the kitchen, each glad to see the other, to escape the pressures that had grown on each of them. Ruth worried about Joe, Grace about Dick, but neither knew each other well enough yet to pry in those directions. Grace was sure anyway that Ruth would cope, Joe was a good husband and father, faithful and caring. She wondered why she'd thought of the word faithful. It was a hidden thought that was surfacing more and more nowadays. She was fairly sure Ruth didn't have those problems.

The boys were in the front room, idly chattering amongst themselves. On Joe's orders they were watching the evening "homework" session on PBS. With no homework now because exercise books had become things of the past the public service channel had launched out into education in the evenings. It did little for their ratings but had given them a cast iron excuse to get some more Government funds. Joe had played a large hand in that, but his boys didn't thank him for it.

The second the session finished the switch was turned over to a Western. In the kitchen Ruth laughed "It's like a time signal these days."

Idly she wondered if Joe could get home to eat tonight. So often now he didn't, staying later and later in the office as the crisis wore on. Grace asked how she coped in the house. Ruth laughed, spread her hands wide and explained "It's just as well we get used to the odd fast now and then. It kinda gets us in practice. But we cook smaller portions now -- just so we don't get to throw anything away. My Joe, he's got this thing about waste now; keeps telling the boys to think of the people in Indonesia." Grace winced mentally. She spent a fair time thinking about the people there too.

Ruth realised what she had said and joked "At least Joe can get home nights. I bet Dick wishes he could too." Grace smiled quietly and hoped she was right.

Ruth changed the subject. "I tell you what gets to me most, apart from the obvious things like not having toilet paper; it's not having tissues and things to wipe away face cream. It's just such a normal reflex. That, and not getting my local paper each week. They won't post it out of town anymore. I used to love reading t. OK, I can get it emailed to me but it's not the same. And when Joe and I sit down in the evening with a drink, I'd sometimes like a cigarette with the newspaper after dinner. Have you tried those ladies' cigars yet? There is just no way they're feminine. The boys saw me with one once and they're like Joe, they don't really like smoking anyway and they really made me feel like I was some sort of addict. But I must say I really did enjoy a cigarette."

Grace smiled. The thought of Ruth with a cheroot was one to be savoured. Grace sighed. "You know for me I think the very worst thing of all is not being able to get my favourite perfume. It was actually called *Wood* Nymph and you can guess how long that lasted. Dick loved it too. And I hate this business now of going along to supermarkets with jars and tins all the time and having stuff poured into them, instead of buying it all in packets. Crazy isn't it -- I never thought I'd miss cardboard so much. Most times when I get home from those shops I feel like a pack horse."

Ruth agreed enthusiastically. "Thank God I have Joe and the boys. Just once a week I make sure I have them all around and they come with me to get the stuff. Makes them appreciate what's going on -- particularly Joe. You can get too remote in that White House you know."

She hammered heavily into the steak, turning to Grace as she did, showing her the wooden hammer. "Joe even wanted me to turn this in for re-cycling. I told him no way. It was my mother's and I don't even know where she got it from. I told him they'd have to fight me for it, but if he had his way he'd have everybody re-cycling everything, even their furniture. Then I told him that if we did that we might end up with single beds and that soon cured him. Not that he sleeps so well these days."

Ruth wondered, as she said it, if Dick slept so well. She glanced sideways at Grace. She was certainly a beauty. Surely that Walton man must be mad, to spend so much time away. She'd ask Joe if he shouldn't be home more. She was a nice girl this Grace, she needed a man and she'd obviously had her heart set on Dick. She agreed with herself. She knew she was right. Joe should see to it.

Grace was talking to her, she'd only caught part of it in her reverie. "I said how do the boys manage?"

Ruth smiled warmly. "You know in a crazy sort of way they love it. Somehow it stretches their imagination all the time and it's good too because they know now that their Papa was right all the time. So, when the other boys at school complain at something the President does and Joe gets the blame too, the boys can turn around, they can be proud and they can tell them their Dad was right, that if they'd all listened to him none of this would have happened. I think that makes them feel good."

Grace envied Ruth's simple pride in her man. She hoped one day she'd be the same. She felt she had so much to give. She felt insulted too. She was as bright as Dick, had as much to do as him, yet everything always depended on his whim. Maybe that was true for Ruth as well, but at least she had her man around, and he appreciated her.

The Western on TV ended and almost simultaneously Joe arrived home. Within twenty minutes they were all round the table, a buzz of chatter and occasional laughter too. Joe asked if Grace had heard Bob's latest piece from Detroit. She shook her head. He told her "It was very good -- he really got down to the heart of things there. It sounds like he's finding plenty to report. And a lot of it's nasty too."

He suddenly looked troubled. Ruth wondered what it was, and the boys just asked outright. Joe paused. "It was just a nasty incident he reported, but it kinda got to me."

"What was it" they asked, ghoulish as only boys could be. Joe looked over at Ruth, apologetic now that he'd mentioned it. He told them of the gang that had broken into the synagogue, broken the old rabbi's arm as he'd clung on to his scroll and how the old man had died. The table was silent. Tears stung Grace's eyes. Ruth swallowed hard. Joe apologised. "I'm sorry. I shouldn't have mentioned it."

The boys had more questions. "Did they get anyone for it?" Joe shook his head, sick now at the thought. "I don't know. It didn't say." Ruth broke her silence.

"But the scroll. How could they? And it couldn't have been any use to them."

"It was parchment" said Joe. "Now let's change the subject." He turned to the boys. "How was that PBS stuff tonight?"

"Boring" they replied in chorus. Ruth laughed, the atmosphere eased, and the boys talked about the Western. "I guess you didn't learn too much from that" smiled Joe.

"We did too" said the eldest. "You ever really looked at a Western?" Joe confessed he hadn't lately. The boy was full of enthusiasm. "I sat watching it this evening and you know it's on the prairies. Well, I kind of wondered if that's the way the whole of the country is going to look soon. Maybe they could get the PBS to show Westerns for educational purposes now and then. Give people an idea of what it's going to be like."

Joe laughed out loud, the ingenuity of the young. But the boy was serious.

"They take a lot of trouble getting those films right and if you look at them carefully you can see how they managed. They didn't have a lot of paper and yet they got by. Maybe we could learn something from them?" the boy mused.

"But they cooked their food on wood fires -- didn't they."

"Yeah, they did" replied the boy "but we don't have that problem, so we don't have to worry about that. We're OK "

"But what if you have to use firewood?" asked Joe.

The boys, in chorus, pointed out that nobody these days had that problem.

"Not here" agreed Joe "but what about say Indonesia where Dick is?"

The boys fell silent. the meal was finished. Joe waited until the plates were cleared away.

"Now we've finished eating I can tell you what happens in a lot of places in the world. The people light their fires with dried up dung from their animals." The boys grimaced.

"OK" said Joe "it's not nice to think about maybe, but you know there's a problem in that too. That dung should be used for fertilising the ground, to get crops, but they don't have the chance in the dry parts of the world. They need to heat their pots that day, they can't wait for the plants to grow. If they could they'd have more food. It isn't easy."

Ruth bustled back. "Joe Simon -- they've finished school for the day -- and you've finished work. Grace doesn't want to listen to all this. What kinda family is this? Come to dinner we say, and we'll talk about death and dung. I should have such a family."They smiled and trooped into the kitchen to help her wash up. Ruth shooed Grace and Joe back into the lounge with coffee. Joe, away now from the family, still looked troubled.

Grace gently asked what it was. Joe sighed. "It doesn't get any better. We had the preliminary report from the weather bureau today. It doesn't look good. We'll get confirmation pretty soon now when they've done some more checks, but it looks like our worries about the climate could be right."

"You mean the heat's going up"? asked Grace.

Joe nodded glumly. "We could be wrong, but I doubt it. I need to call Dick about it too. Can you remind me in the morning"?

She agreed she would. She'd have a genuine reason to find out what Dick was doing. At least it would be a reason, not just an excuse this time.

## CHAPTER ELEVEN

Lestari, sitting at Dick's feet with her head resting on his leg, turned to look up at him. She smiled. He always looked so relaxed after dinner. Softly he stroked her hair. From where he sat he could see the soft alluring curve of her breast beneath the white lace of her blouse. He smiled to himself -- she never wore a bra in the evening. He pulled himself up and slid his hands down over her shoulders.

The phone rang. He cursed, but took it anyway. Joe Simon sounded grim, and, in sparse outline, went through the initial report from the weather bureau. For late spring it was already unusually warm and now the long-term forecast spelled out more of the same. Walton knew what he feared, a repetition of the climatic changes that South America suffered after deforestation.

Lestari was forgotten now, and Walton agreed to check with the Indonesian weather service next day about long-term comparisons. He wanted to be sure the latest hot spell there wasn't a freak.

"They've had one there too?" asked Simon. Walton confirmed it.

"Dick, there's something else you should check on. It's may be that the timber companies here are panicking, or maybe trying to pull a fast one on us, but they say they've had some problems with the timber that was affected by the PWN. I don't understand it fully, but it has something to do with fibre strength. Does that mean anything to you?"

Walton groaned inwardly. "Joe I was going to call you about that -- I didn't realise you'd be in the office this early. Again, I thought it might just be a local problem here. No, I haven't got to the bottom of it yet and we still have a few checks to complete, but the way it's looking, the timber is almost useless other than for firewood. Here that's not so bad, because it's what most people use it for -- other than exports of course -- but if it's the same in the States then we really have hit even more trouble."

"You mean none of the stocks could be used at all?" Joe sounded horrified.

"Joe, I'm not sure of it yet. I need to do some more tests here first, then compare them with the ones they've done in the States."

"Dick that sounds like the end if it's true. You know I run this allocation committee that sorts out the priorities for who has what. Well we'd been estimating that with all the wood that's been cut down we had a sort of bank that we could draw on that would last us for at least a couple of years or so and we'd then be able to stoke up that with at least some paper from stuff like papyrus and bagasse and of course re-cycling. But if what you think is true then all that wood is gone. It means our bank has an empty vault."

Walton interrupted. "Joe don't panic. Maybe there's a way of treating the wood, maybe we can stiffen up the fibres somehow -- there's no telling what we might be able to do. You'll just have to give me another day or so and then I promise I'll be back to check on your end of it. OK?"

"Dick I'm going to have to warn the President. John's got to have some idea of what's going on. This could affect his whole international strategy. When will you be back?"

Walton paused and thought for a second "Say 48 hours from now."

Joe was relieved. "Hey, Grace, will be real pleased. I'll get her to call you."

Walton moved the phone slightly away. "Joe that's nice of you but I'll check in as soon as I can -- I've got a busy time ahead." Simon wished him luck and asked him to report to his office as soon as he arrived back.

Walton slumped back in the chair. The nagging current of fear which had haunted him for the last two days had suddenly become reality. The news about the weather was bad enough, but that was bound to take time before it had any effect. But this? This was the worst news yet. Lestari had been by his side. She'd heard it all.

"Do you think it could happen Dick?"

He held her. "I don't know, it's not my field. That's why I'm going to have to go back tomorrow night and start work with the people who do know. And I have to look at the other results they've got for me,"

She looked sad "It sounds as if you'll be away for a long time Dick -- will you?"

He was not much happier than she was "I don't know. I hope not. I couldn't be without you for too long now."

She raised her eyebrows. “Dick, that sounded very much like a compliment. Are you alright?"

He laughed suddenly "I think I need a thorough medical check-up. Now in my favourite fantasy the nurse has sexy black stockings and... shall I go on?"

She stopped him, put her finger on his lips, took his hand and put it on her knee. "I think you will find this nurse forgot a lot of things." Walton investigated. She was right.

Lestari smiled. "I think the nurse’s job is to undress the patient don't you?"

He grinned his agreement and stood up. She teased him now "I think I detect a swelling here."

He couldn't wait any longer. He picked her up, carried her to the bedroom and laughing together they fell on the bed. It was, thought Walton, the only way to say goodbye.

***

He caught the scheduled Garuda flight. As usual there was a lot of curiosity. Only rarely now did the hostesses see a man carrying a briefcase, even in the first-class section. He noticed their attention and thought again how women were turned on by power. He wondered if Grace was still subject to that.

He thought ahead to how he was going to handle that. It would have to be a special effort. This time he must try to be fresh and rested; this time he must make an effort with the charm. OK so she was a bitch sometimes, but if he was going to be in America for some time he couldn't afford to fall out with her. He didn't need the fuss.

Dejectedly he studied his notes. If the heat threat did materialise it wouldn't be just a matter for world concern about a recession and exports, it would be a threat to a whole series of coastal cities. He quickly dismissed the thought from his mind, that was Lean's problem. But the thought stayed with him awhile. He worked out some equations in his mind, and inwardly sighed at the answer. He'd need to check it but according to all the evidence, a rise of just 1 degree centigrade in average temperature could mean as much as an 11 per cent loss in grain harvests in middle America.

They usually had a surplus in most years, but now with the heat threat, it was bound to mean more losses in other parts of the country through erosion. At the very best it could mean shortages of bread. But would that be so bad? Maybe even people would be better for it. Already he'd heard reports that there was a boom in cycling in the major cities and with the decrease in the use of cars people had to be walking more. So maybe it wasn't all bad after all. He was sure Joe Simon would approve.

***

Joe Simon, next day, was in no mood for such trivialities. Somewhere in the back of his manner, Walton felt, there was some sort of resentment, a feeling that Walton had been away from the sharp end, enjoying himself maybe?

They exchanged their notes and solemnly read them. Each was almost a carbon copy in its content. In each country the fibres of the wood were soft, not having the strength necessary for any but the most basic of papers. And, similarly, the heat seemed like it was increasing. Joe finished first and looked up. Walton felt his gaze on him, could feel him expecting a comment. He looked at Joe, then silently shook his head. "What is there to say?"

Briefly he told him about his calculations concerning the grain harvest. Joe groaned out loud this time. "Yeah we've had much the same sort of estimate already. Just how much more are we supposed to take?"

Walton helped himself to a coffee and poured Joe one. "Can you get any rawer material into that pulp bank of yours?" Joe shook his head.

"Not a lot. Mind you we are getting more ruthless about it now."

He outlined the President's latest emergency measure. All phone books were now to be handed in, all fencing was to be commandeered and even libraries were to be 'rationalised.'"

"What's that mean?" asked Walton.

Joe smiled "It means that each city sits down and sorts out which ten per cent of its books it wants to keep. Even universities. The whole range, no matter what it is."

Walton was appalled. "But you could be destroying whole bunches of knowledge we might need."

Joe agreed. "But it's new knowledge we need now Dick. We need solutions."

The President had swooped in other places too. All state and government records were to be ruthlessly swept up, cleared out and put into a central dump in each city. Tax records, health documents, commercial filing systems too.

Joe explained "You wouldn't believe how much of it there is around you know. Even families are going to have to go through their own books, their own kid's comics, everything." He added "There will be appeal panels of course, but they've got to be told to be very tough. We reckon people ought to have an allowance of one religious book and say six others. They can always swap them amongst themselves. At least that way we have an emergency stock for re-cycling."

Walton broke in "But you do know that's only temporary don't you? I mean you can only re-cycle stuff so many times. After that it sort of wears out, and in any case re-cycling only works for some grades of paper."

Joe agreed, but explained "Dick, you don't seem to realise how serious things are. The whole of commerce and industry is on its knees. Those that can pay taxes can hardly be forced to now and in a lot of places the people that are working are in a minority. It's only a matter of time before we don't have any more money to hand out in welfare payments. Then what happens?"

Walton didn't know. And he guessed nobody did. Another thought struck him. "Did you get any supplies of pharmaceuticals? It was only when I was in Indonesia I realised we get practically all our quinine from there."

Joe interrupted him. "It's a lot wider than that. There're things like cortisone and a whole mess of other drugs. I never knew so many of them came from trees. I don't think anybody did."

"But are you getting any?" asked Walton.

Joe smiled ruefully. "We're getting some. Would you believe the World Health Organisation has decided to treat us as a nation in need of world aid -- a poor country? Imagine? The US of A being classified that way."

Walton stood up, investigated the coffee and found it empty. "I've been hearing Bob Powell reporting from Detroit. It sounds like there's a big crime problem there."

Joe sounded disgusted. "It's all over Dick. It's everywhere. A lot of the old Daley crew want us to pull some troops back from Europe and they're calling for martial law. You know one of them down in Alabama is actually pressing for the death penalty for wood theft?"

Walton laughed "That's par for the course. Fat chance."

Joe stopped him. "Dick it's likely to happen. People are getting that desperate."

"But aren't there any good sides to it?" asked Walton. He sounded almost desperate now.

Joe phoned for some more coffee and relaxed slightly.

"Yes, in all sorts of trivial little ways there are. There's religion of course, that's making a comeback and there are groups of people really setting out to help each other. And, here and there, people are getting fitter. There's a lot less people turning up at doctors' offices nowadays. Maybe they can't afford to, but my impression is they're eating less and getting more exercise. And a lot of the unions have been coming up with good ideas to get their people jobs. Things like factories switching over, maybe as co-operatives to get into things like solar energy panels. It's maybe overdue perhaps, but gradually that sort of thing is starting to happen. Then there's the way people are watching less TV, though that's only because the networks have had to cut down their hours due to the lack of commercials. I mean who's got anything to sell?"

Walton looked more cheerful now. And Joe was happier too. "The other thing that's been happening a lot is whole communities getting together. One group growing one crop and one group another so that they can barter amongst themselves to cut down on food bills. Maybe it's wishful thinking, but the country's kind of leaner and fitter in some ways. And of course, nobody lays around reading magazines anymore because they just don't exist."

"Even the girlie ones?" joked Walton.

"Even those" grinned Joe, "though I was offered an old one for twenty dollars the other day on the street."

"What did you do?" laughed Walton.

Joe smiled "I told him who I was and asked him to send me a bill."

The coffee arrived and this time Joe poured it. He looked at his watch, then at Walton. "I guess you need some sleep. Why don't you go home rest and get back in tomorrow? There's a meeting of the President's scientific committee and that could be sort of crucial."

Walton gulped down his coffee, asked about Ruth and the boys and, as he'd hoped, Joe invited him and Grace round for a meal. Joe teased him "I think you'd better go and find that young lady before she breaks the door down."

"Or over my head?" smiled Walton.

"That bad huh?"

"I hope not" said Walton, but all the same he had a feeling the next half hour could be difficult. If only he'd had the time to buy her a present. But then what chance had he to go shopping. Lestari had been with him right to the last moment. And he couldn’t explain that to Grace.

He knocked at her office door. "Come in -- who is it" she called. He gave a muffled reply and knocked again. She came to the door, opened it and stood there.

The outline against the light was like a fashion silhouette. A slim waist flaring out to trim hips and -- he noticed instantly she was wearing stockings. For a second there was an awkward silence. He joked it away. "I just thought I'd drop by on my way through."

She ushered him in, he pushed the door shut and held her. She thrust against him, all defenses down now; just so glad to hold him again. Walton relaxed slightly. It was a good start. He made a mental note to carry on this way.

She pulled away from him, put her hands on his shoulders and studied him closely. She flicked a hand to his ears. "You know you have a grey hair coming through?"

Walton smiled "I looked it up once -- apparently it's classic sign of sexual deprivation. I think it needs treatment."

She pushed him away now. "I would prescribe some rest first. I’ll make a house call as soon as I'm finished here."

Walton looked puzzled for a second. "Excuse me, but aren't I supposed to be the boss around here. Has something changed?"

She giggled. "Oh Mr. Walton didn't you know. I'm now the secretary to the President's new Scientific Committee. I guess there's not too much paperwork, but it is sort of important. And there's a meeting tomorrow."

Walton coughed. "I do know. I happen to be going to it."

She came to him again "Then we’ll need an early night won’t we?" Walton felt better now. All the old rapport was back. Somehow too it was better that she wasn't working just for him. It gave her more status and independence and that had always suited her. There was something about women with confidence. He teased her again. "Since I'm a member of your committee I absolutely insist on a briefing beforehand. Say about four hours' time at my place?" She looked at her watch. "I guess I could probably make that" she smiled.

Now she held him and this time there was a touch of urgency about her. "Dick do you know how I've missed you?"

Walton muffled into her shoulder. "Missed you too, though maybe 'til now I just hadn't known how much. I'm OK mostly during the day, but the nights, of boy. And then I'm OK 'til I see you again, then it all comes back."

"If this is what jet lag does for you, you can go away any time Dick Walton" said Grace. She was all softness now.

He hugged her and confessed. "But honey I have to get some sleep now. Do you mind?"

She bustled him away, confirming the date. She said softly "Don't start without me."

***

The car was quicker through the streets than normal, less hindered now there were fewer cars on the road. Walton noticed the horse on the sidewalk. The chauffeur told him it was the latest fashion. He lugged his cases into the flat, shut the door and kicked open the post box inside. It was empty. What had he expected? As he put away his clothes, dropping them into the laundry cupboard, he flicked on the radio. The forecasts for the grain harvest were gloomy and this time it wasn't just the farmer's traditional tale of woe. The best estimate was that the harvest would be down by around 15% -- more or less what they'd hoped to export for badly needed foreign currency. Their dilemma was simple. If they did export then there would be shortages inside America. It had to be a government decision.

The bulletin ended with a new specially detailed forecast for the increasing number of amateur small holders. A few adverts for seeds, then a new programme, hints for people setting out to turn their gardens into vegetable producing areas.

People had after all more space now, with all the trees gone; and with welfare payments drying up it was almost a necessity to grow at least some of your own food. Walton thought back to the old days of

TV dinners. Now they'd all gone too, vanished the way of record covers, receipts, cheque books and the rest. He slumped into bed and in minutes was out cold. He was woken by something warm.

It was a breast. Grace's. She was on her knees by the side of the bed, just looking at him. He yawned, stretched and she slipped the covers back, studying him as he lay there on his back. Suddenly he felt guilty about the tan, she'd noticed that before, but how could he hide it?

"For an old man with grey hairs you look to be in pretty good shape. In my opinion I think you'll probably last the night."

Walton studied her. What a sight? Who needed girlie magazines with this sort of thing around? He gave out with a mock groan of pain. "I can't make it alone though."

Grace slid in beside him. When he woke up again he was smiling. It was good to be home.

## CHAPTER TWELVE

Walton watched as Grace poured the cereals from the plastic container.

"You know" he observed "I could get withdrawal symptoms. I used to collect the cards in corn flake packets. Did you know that?

Grace giggled. "You're a fool." "And you know what? You realise if this goes on there will be couples all over America talking to each other over breakfast. No newspapers, no funnies. It could be disastrous."

"I can't imagine how my father's coping" said Grace. "I don't think my mother's seen him over breakfast for the last thirty odd years."

"But" said Walton "there is still the good old radio. Let's see what's new." Grace hesitated for a second. Powell had called her and asked her to keep an ear out for this morning's background piece. She switched the radio on and hoped Dick wouldn't mind.

Walton nodded occasionally as the reports came in from around the world. NATO leaders were getting together so they might be able to help out America with the costs of keeping troops abroad. Lean had told NATO bluntly that they either met the bill or he'd find jobs for the troops at home. He just couldn't afford the tab anymore. Walton nodded in agreement. It was about time Europe started looking after itself.

How would they have managed in Bosnia and the Middle East without the Americans? The dollar was still being gloomy. Walton took another mouthful and wondered how much longer they'd have crunchy flakes.

Sales of cigars for women were eventually catching on -- some bright guy had found a way of colouring them. The weather again was going to serve up an above normal temperature. And traffic conditions were still light. He looked over at Grace.

"Had enough?"

She hesitated for a second "No let's keep it on -- there's sometimes a background piece after it."

She felt guilty now -- she should have told him straight out it was Bob's star piece of the day. He was doing well -- Voice of America was picking up almost every piece and that meant extra money for him. Walton didn't seem interested and was toying with his coffee. Then the announcer linked into the Powell special. Walton was suddenly alert, wondering in the back of his mind if Grace had known about it. He had his suspicions about that guy -- though perhaps he couldn't blame him too much. Boy, he remembered, had she been sensational last night. Really erotic. And then this morning too...it was like she just couldn't get enough of him.

Powell's piece started and he noticed Grace stiffen slightly. He couldn't resist teasing her and she smiled back.

Powell was in Detroit and was telling how he'd met a guy in the line at the bank. Name of Joe Williams, worked on the track at Chrysler, unlucky for him he'd been a victim back in '79 when Hamtramck closed, but that wasn't so unusual. He just joined those from Trenton and the other Chrysler plants 'til it picked up in '83. Now he was laid off again.

Right now, Joe was OK --just about, there were still welfare payments. He and his wife Nancy had managed to pay for their house way back in the good days and raised their one boy. Joe hadn't wanted him on the tracks and had got him a job as a car salesman. That had been fine, except there was no call for Cadillacs just now. So, the boy, who had two boys of his own now, was also on the welfare line. He had no union benefits to help him through.

Powell sounded sort of sad, yet beneath it all was the anger reflecting through old Joe's voice. "I only ever had that one skill" said Joe. "I guess when the welfare money goes, I could be in trouble. But better than some. I have my garden and you'd be surprised just how much food you can get from that when you really work at it. Never did before now -- just kept it as lawn for the grandsons to play on. And I can go fishing on Lake St. Clair with my son; that can help too, and Nancy cooks them wall-eyes real good. She don't have a car now of course -- we traded that in like most folks 'round here. Not that it fetched much, who wants one of those big estate wagons these days?"

Had he thought of selling up? Powell asked. Joe was thoughtful for a moment. "No but my boy has. He's heard there's some good jobs going in Brazil. Seems like that's the place to be and lots of folks are thinking of moving away now. He sounded suddenly sad. "I guess" he said "the way they say this thing is going, by the time it's over the house might be my boy's anyway by then if it's OK to come back. Who can tell?"

Powell probed gently again. "You had a visitor the other day Joe?"

"You mean the guy from city hall? Yeah I did and that's something I ain't going to forget for the rest of my life."

Over the years, Powell explained, Joe had built up a valuable collection of books on the motor industry. They lay cherished in his study, he pored over them and had his friends come around to talk about old times.

The visitor was a State inspector, making sure each house had no surplus books. One Bible and six others, that was the limit. All others had to be taken away.

Joe resumed his story. "The guy was sad about it too -- got real interested in the collection, said it should have been in the library in ordinary times.

"'Cept there hardly is one now. I kept the bible, and two of the books. We kept one of Nancy's favourite cookery books, the one that came from her mother and I had to keep one about gardening since I don't know too much about that yet. She wanted to keep a hymn book too -- she goes to church more now you know -- and that left us just one. That last one was a problem."

"What did you keep?" asked Powell.

Joe laughed a trifle bitterly. "It was a world atlas. It might come in useful for the boy and who knows, maybe we'll have to move on soon."

"I guess these days you have to be practical don't you?" Powell mused.

Joe laughed and added "My boy sure is. Selling Cadillacs of course wasn't easy, but he had one for himself. He kinda had to didn't he? Well when the crunch came, before they sold the wife's car, he tried real hard to trade it in. Nobody wanted to know about it. You know what he did? He even tried to barter it for a cow. But the man turned him down, reckoned the Cadillac would only make a down payment and my boy knew he couldn't keep up the payments on the cow. You know I often kinda wonder what old man Ford would have made of all this."

Powell had one last question. "And how's your boy now Joe?"

Joe sounded sad again. "He ain't well. You know he fought for America. Went out to Nam and all he got for it was malaria. Well that was fine until lately. He just took his quinine when it got real bad and then he was OK. But they ain't got none of that stuff now. Not even for the guys who went to Nam. Now don't that strike you as real bad?"

Powell summarised his piece. "Right now, in Detroit that's not the only thing that's real bad. Factories look like dinosaurs dying from starvation, no chemicals as raw materials to feed on. The weight they shed is human, regular guys like Joe Williams and his son, through and through Americans who've done nothing to deserve it. So, they feel bitter. And, yes, they're looking to Washington to do something about it -- though even they don't know what. They're bitter, they're confused and some, now downtown, are getting hungry. More and more they're looking to up and go. This country's always been one that's on the move and in the end a town that doesn't give them jobs is going to lose them. But just where to go; where is it any better?" Powell's newscast ended.

"It's a good question" mused Walton, half to himself.

Grace nodded silently. "That poor man Joe -- and his son too."

Walton looked over at her. "That guy's a good reporter -- got a lot of heart. I just didn't know it was like that."

Grace said nothing. Walton studied her for a second. He really cared for this girl.

A frown flicked across his brow. "You know it's going to get to us too. Up until now we've been the lucky ones -- working at the White House and all that, but in the end it'll catch us. You know soon there won't be much reason for a lot of people to live in Washington at all. And it could get violent too. Can you cope with all that?"

Grace put her hand across the table to his. "I think that's one of the most charming speeches I've ever heard you make Dick Walton. Yes, I'll manage. And if I can't then I'll just get in the car and drive back home to my dear old folks. They've been dying to see me anyway. And they've got enough land to start cornering the market in crunchy flakes any day. You should see it sometime."

Instinctively Walton stiffened. Home to see the folks? It had that ring about it. He joked it away. "Didn't you know it was only your money I was after?"

She came over to him, grabbed his hand and held it to her. "If that is so then you must think I keep my money in some very peculiar places!"

She wheeled away. "Hey, do you realise we are very, very late indeed?" Walton was in the other room now.

"That's all your fault - you and your fetish about showers." She giggled. Ten minutes later they were on their way.

Her compact zipped through what traffic there was, and she tore into the car park. Walton uncovered his eyes, breathed a mock sigh of relief and undid his seat belt. "Now I know why people go to church." Half an hour later they were two different people. Walton was the scientist, she the secretary. Around the long table, with Joe at the head, were ranged some of the best scientific brains in America. Walton felt flattered to be amongst them.

The talk was of alternatives, almost all now accepting that a solution in the short term was unlikely. Joe had explained it in opening the meeting.

"We have to face it -- whatever happens now we are unlikely to get a normal crop of timber for many years to come. Leaving aside the climatic consequences for the moment -and I know they could be horrific too -- let's see if anyone has any workable ideas for a wood substitute. Dick?"

Walton coughed nervously. "I think sir that the most promising area lies in substitutes for what we need for paper. And they're fairly plentiful. My colleagues will know about bagasses, it's left-overs from crushed sugar cane. Now if you use that and bamboo then we could be like China; they get half their pulp that way. Now as of this moment America only gets one per cent of its pulp from substitutes. But we could get far more. It's just a matter of application. If you want to make boards for instance well then you can do it from corn cobs, coffee grounds, stalks, peanut shells, sunflower seed hulls and a lot more. You can also use coconut trees if you can get hold of them -- but I guess it would mean buying -- and then there's a plant called kenaf. So why haven't we done it already? I'll tell you. The reason is that it's been easier for the commercial timber companies to just go on chopping down trees. That's what they're used to and that's what they're geared to." He paused "Overall the best bet is to do some sort of deal with China for the time being and get hold of some of that pulp."

Joe didn't want to rebuke him publicly, but he had to steer Walton away from that tack. He knew about Lean's deal with Lentov, that as an exchange for no trouble they wouldn't do any deals with China. Joe changed the subject slightly. "What about re-cycling?"

Walton, slightly puzzled that Joe had rushed off the China alternative, carried on. "We use about 70 million tons, but we only recycled about 16 million. What it comes down to is this. At the moment we throw away around three pine trees a year each. Putting it another way we're wasting an area the size of Delaware each year. Had this been a year ago I'd have said that if we saved just ten per cent more we'd be a lot better off, but what we're seeing now is the law of diminishing return. What paper do we re-cycle? Since we don't have any … and I say this with respect, I don't think that's the answer." Walton hesitated. It sounded a bit disrespectful. But Joe didn't seem to mind.

Simon breathed a sigh of relief. He'd steered Walton and the rest away from the subject of China. It wasn't something he could have explained. That was a purely private deal between Lean and Lentov.

One of the scientists chipped in. "There is one problem we've all forgotten. A lot of these plants are fine, and you can get pulp from them in the end. But a lot of them won't grow in the areas we're talking about. So, it isn't a wholesale solution by any means."

Joe nodded "Anybody have any positive ideas?"

A grey bearded old man halfway down the table put up his hand. Joe nodded. "Professor Thorenson." The professor gulped. He wasn't a man for meetings. He preferred wandering through forests, just thinking. He was hesitant. "You know the Americans are a peculiar people, they're always a lot more enthusiastic when they can see an end result for themselves. Like money? There is one idea that has been tried elsewhere and maybe it's worth trying here. But it's long term and it involves growing trees. It's not an instant solution."

Joe was patient with him. He smiled gently. "I do know you don't get trees overnight." The professor relaxed. Then carried on. "They tried it in the Philippines, and it looks like it could work. They got together with the World Bank and they got their forest farmers there to pay rent for some derelict land. So maybe it doesn't sound like such a good deal. But you give a guy twenty acres or so and you tell him that four are for him for livestock and food and the other sixteen he has to have trees on. Now in ten years or so those trees will be his to sell -- but he can only sell them to the government. The government meanwhile gives him a loan, around three quarters of the cost. They reckon at the end of eight to ten years the farmer has had all his food for nothing, he's raised some trees that they need anyway, and he's had a small income on top. It seems like it's working. Maybe they don't drive around in big cars and OK, so they don't get to have expensive holidays, but they seem to like it. At least they live."

Joe was fascinated by the idea. And Walton chipped in too. "I'd heard of that, but I didn't know yet that it worked."

The professor nodded into his beard. "I've walked around some of the places. And you know what the Philippines used to be like."

Joe broke in "What was it like?" He looked at Walton.

"The satellite pictures show it up best. They probably have around 30 per cent left. The government reckons they need 46 per cent. They've just been into the whole thing. The current estimate is that if they carry on the way they're doing -- outside the Professor's schemes that is -- then all the original forests will soon be gone. They wouldn't be able to get enough from the second growth forests even for their own needs and already they've had lots of destruction from flooding and the silting up that goes with it. But they're not alone. Thailand's about the same."

The professor nodded. "It's getting vicious there. The last estimate I had was that around 30 to forty forest guards a year are getting killed. They have gun battles with tree poachers. Malaysia's not much better --they estimate they only have about 12 years production left. We're not alone you know."

Simon nodded, appalled by the scale of the problem. "So, what happens when it's all gone?" he asked of nobody in general. Walton laughed slightly bitterly. "I guess they'll discover Africa."

Joe brought the meeting back to order. "This Philippines idea. Has anybody else tried it? Could it work here?"

He turned to an eager, younger man on his left-hand side. He introduced himself as being from Earthwatch, Joe's old organisation. "Can I just say, without offence, that part of the problem we have is the traditional one of the foresters themselves -- that only they know what is good for the country. Now perhaps things are changing and there are some signs that where the forester is enlightened and recruits the rest of the community he can bring about some positive results. What is quite often needed is for the forests to be given to the people and not to the multinational companies. At the end of the day the companies could just as easily switch their funds to some other product, but the people will always need the trees for themselves."

Joe jumped back in, "OK so how would it work? I'II bet you didn't know it was already underway in Oakland in California. Well there's a group of people there reckon they can grow wood products on vacant lots. Idealistic maybe. It's typically Californian, but in principle the idea is the same -- it's forests for the people. Now maybe too many people here haven't heard of Gujarat. But that place -- it's in India -- pioneered a thing called social forestry."

The gentleman from Earthwatch offered, "They went around the schools, they talked to people, they convinced them that trees were important. I guess we could skip that step here now. But what they did next was to plant trees in all sorts of unusual places. Anywhere that was spare. That gave the people some idea of what could be done. Having done that, they then started giving people seeds and advice on how to plant them. School kids loved the idea, they went very big on it. Soon the foresters went back to the village leaders and talked them into putting aside around ten acres of spare land. They gave the village the seeds, even paid some people to plant them and gave the community the right to graze their animals there. When the trees grew the foresters and the village split the profits. Simple maybe, but it worked."

Joe looked around "Any other examples? Dick?"

Walton looked round the room, estimating the age of those there. "I guess a few people here remember Korea. Perhaps they'll never forget what it was like. Just a succession of brown bare clumps called hills. Nothing on them at all. You go there now and the whole damn lot is covered with pines. I know that's not ideal, but it's better than nothing. I even saw a billboard there *"Love Trees. Love your Country."* Did you know they even have a day devoted to trees? They've got things called VFA's -- Village Forestry Associations and everybody, and I mean everybody has to be a member. What happens is that the association, that's the village really, produces the trees and the wood is split around the village. Any money that comes from surplus timber goes into a fund for other village projects. Now they just don't have a fuel-wood problem at all."

Joe had a question. "But what if the land they need is privately owned?"

Walton grinned "The guy is offered a deal. He's told he either has to reforest it himself, which can cost him a lot of money or he can turn it over for nothing and take ten per cent of the proceeds. Now since he's probably not had a penny off the land for years and he either can't or won't spend the money on reforestation it makes sense for him to sit back, let the other people do the work and take his ten per cent. I ask you who loses?"

Joe grinned. "You know sometimes Dick you sound more Jewish than I do."

He looked round the table. "Can I summarise it then? We have two courses of action as I see it. One we get on with planting what sort of paper bearing substitutes we can. That's for the short term and would maybe get us some paper only by around this time next year. Not enough as we've been used to, I know, but enough perhaps for some essential supplies. Next perhaps is this idea, rather longer term, of trying some developments along the lines of social or community forestry. Meanwhile we try to crack the PWN problem, though I guess in time they'll die of starvation anyway."

He concluded "I know we all have at the back of our minds the climatic consequences and the problems that are going to come with erosion. Just to give you an overview I have to say there are some fairly alarming reports of dams silting up at an astonishing rate. I know it's happened elsewhere, but the suddenness of all this could bring on the effects fairly sharply here. I guess we can all expect some shortages of electricity before we're through. On the climate front the immediate problem is going to be the next grain harvest. Overall it looks like we have a choice between keeping all we grow and having just about enough for ourselves or going short here in order to get some foreign currency. I shall be recommending to the President that we impose some fairly minimal sort of rationing here in order that we can afford to buy in some pretty desperately needed goods, and I'm thinking mostly of medicines and some raw materials to try to help what industry we can. It may sound simplistic to you gentlemen, but I'll forego the bun on my hamburger if it means we can import some drugs and maybe some lignin and cellulose too."

The meeting broke up and Walton stayed to chat with the Earthwatch delegate. They had a lot in common. Grace hovered in the background. Joe strolled over, repeating the invitation for her and Dick to join them for supper.

Walton heard them, said goodbye to the Earthwatch man and joined them. They were chatting about Powell's piece from Detroit. Joe had been looking at the emigration figures that morning. They were rising. A lot of people were just giving up and going abroad. Joe smiled at them both "I hope you two are going to stay around. I've sort of come to depend on you."

## CHAPTER THIRTEEN

In Kansas, Powell flicked through the slimmest edition yet of The Star. It was full of gloom. Worries over crops and concern, too, at the way the wind was scouring the surface of the fallow land. The paper's leader column was to the point: "Scientists state that unless we get some crops back on those lands we could be in for a repeat of the Great Dust Bowl. Don't those folks in Washington know that takes machines and they need oil? When are they going to realise they depend on the breadbasket of America for their food? Either they give us the oil, or they get no food."

It seemed too that everybody now wanted to be in the mid-West. A feature article told how people in the cities were moving West.

One farmer faced with the problem of how best to use his fallow land and stop erosion put a small ad in the New York Times. He put up the land and a shack for rent, labelled it "the New Pioneer Scheme" -- and was almost bowled over in the rush. One guy jumped on a plane, took a quick look and offered him the price of his apartment. The farmer took it and now the New Yorker had set up as a smallholder. The word spread and now the city dwellers were looking for small plots. Real estate agents had picked up the New Pioneer label and a mini boom was underway. The article concluded "The folks in the East call food the new oil -- and they could be right."

Powell rang one of the agents, posing as a buyer. The man was weary. "You done any farming before?" Powell confessed he hadn't. "It ain't easy going you know" warned the man "you get aches in your back and you get blisters and it's a long way from the movies."

Powell then explained he was reporting, not buying. The man laughed "Oh boy, could I tell you some stories about those city folk. You know one guy was out here for two months and he bought himself a couple of cows and read a book about how to milk them. Then when he milked the cow and got no milk he went back to the guy he bought the cow from and complained. He just didn't know about cows being pregnant before they gave milk. He just didn't know."

Powell could almost hear the guy shaking his head. Powell got from him, the name and number of a farmer he could call. The farmer was helpful, but worried. There had been a lot of pressure to produce more corn, and that meant almost constant cropping. "In the old days" he explained "we knew the land had to take its time awhile. We'd rotate it, corn one year, some vegetable or other the next. Now they just want corn all the time, they just tell us *Put on more nitrogen.* But that ain't cheap any more you know. The prices are OK right now, but I don't know how long it can go on that way."

"Why not?" asked Powell, who knew nothing about farming.

"It's like this son, you need good top soil, but the way things are going that's running away fast. There ain't nothing to hold the rain steady in the ground no more with no trees and there's lot of folks say we're just cashing in short term, that real soon the soil's going to get sorta thin and when that happens why it won't grow nothing anyways. See what I mean?" Powell did and asked who the experts were. The farmer mused, "I'd say I knew what I was doing. The farm's raised my family for some years now." Powell apologised, explaining he wanted an overview of the whole corn belt. The farmer gave him the number of a university experimental station in Iowa. The research department was helpful, eager to push its cause.

"Just take the land in this state" said the researcher. "We lose 200 million tons of top soil each year. And that was when we had trees to help out. Now I just don't know what the rate is. You can almost see it dwindling before your eyes."

"Can't it be replaced?"" asked Powell.

The man laughed "Not while we're alive, nor our children for that matter. "Do you know how long nature takes to put down seven or maybe eight inches of topsoil?"

Powell had no idea.

"Guess" said the man.

"A hundred years?" ventured Powell.

The man sighed "I suppose most people in the city must think that. The answer is 7,000 years."

Powell gasped. "I guess it's probably academic but what about erosion rates?"

The man paused "This will surprise you too. If it's in a forest that's left alone then they once worked it out at 174,000 years. Now if it's a meadow it comes down to 29,000 years and if you're really careful and rotate the crop properly then it's down to a hundred years."

"And if you, plant corn all. the time?" asked Powell. The man sounded depressed. "We know the answer to that one -- it's 15 years. And that's what we're doing right now."

Powell groaned. He thanked the man, put down the phone and thought about it. No wonder the man sounded depressed. No wonder the farmer was concerned.

Powell phoned in his piece for the station. He concentrated on the New Pioneers angle and the farmer's tale. Folks would understand that -- the rest he still just couldn't believe. He walked around the downtown area but soon headed out of it. There were so many people around these days.

In the side-street he heard the sound of singing. it was an old French style chapel and the sound of spirituals was loud and clear. He sneaked into the back, exchanged "Praise the Lords" with the old black man on the door and sat with the rest as the preacher started. "My question today is simply this: What would God have to say about all this?" It was, thought Powell, a very good question. The preacher had the answer: "'I tell you what he'd say; 'You down there -- this time you almost went too far.' He'd say 'I gave you the land, I gave you air, you had water and the fruits of the earth. You had trees and you had animals to serve you. And boy… you poisoned them all. They were my gifts and you wasted them.' And I guess God might weep a little over that, but mostly I reckon he'd be kinda angry. And when he's mad he sends that wind, he holds back on the rain, and the land … it just blows away some other place where folks know better. I guess we oughta pray for his almighty blessing and tell him we're just chillun still and that we didn't know no better 'til now; and I guess we should go down on our knees and promise him that this time we'll look after his gifts real well. I know my brethren that it's not our fault and I'm sure that in his wisdom God knows that too. Lastly I guess we should maybe say a prayer -- though I can't see they deserve it -- for those folks who started all this. Let's use the words of our Saviour *'Father forgive them for they know not what they do'.* Hallelujah. Praise the Lord." The congregation rose and sang again, Powell left quietly.

The words "Father forgive them" stuck with him. Kansas and Detroit had that in common: the turning back to religion when times got hard. Maybe it was something about America as a whole; maybe it was to do with the scale of the problem. It was all too big for people to grasp.

He went back to his hotel room. He needed a bath to wash away the sweat of the day. He had never known it so warm here before. He dried himself, changed his clothes and thought about what to do next. If only he had something to read, Powell looked around the room. Nowhere in sight could he find a single printed word -- there wasn't even a Gideon bible. Alone in a hotel bedroom he reckoned you had maybe four choices. You could eat, sleep, screw or talk.

He lifted the phone. Maybe his editor had some gossip;, maybe it was time now to go back home.

***

Walton stirred and felt Grace move against him in the bed. She felt warm and snuggled up closer to him. He moved over onto his back, stretched his arms behind him and thought that life today was pretty good. Last night had been fine too, Friday night dinner at the Simon's -- he felt privileged to have been invited to their Sabbath supper. It had been simple and homey and very relaxing. And it had got to Grace too. He smiled -- Ruth was definitely matchmaking and Grace had confirmed that later. For once it hadn't made Walton feel that uncomfortable.

Looking at Grace across the table she'd looked the picture of modern American womanhood, beautiful, composed and confident. A lady in the lounge and a wench in the bedroom. What more could a man want?

Grace laughed softly beside him. "Dick, you are looking very, very self-satisfied." He turned to her, pulled the sheets over them and started to caress her. Her hand moved down to stroke him. Suddenly she pulled away. Walton was startled. "Hey, what's the matter?"

She laughed. "Dick Walton. It's Saturday and there are things to do." He agreed and pulled her back towards him.

"I didn't mean that", she said. "I meant that today you are going to be the liberated, sharing, caring American male. You're going to help me with the shopping. It's about time you came down to the realities of life."

He slid his hand to her breast. "You're right -- that feels very real to me."

She giggled. "We have all day -- after the shopping. Come on, I've made my mind up. You can start breakfast while I have a shower."

He sounded shocked. "You're going to have a shower alone? That's anti-social, you know?" She ran from the bed-room and he heard the bathroom door lock behind her. She shouted, "I want two eggs easy over with toast and coffee."

He groaned. She meant it.

***

They took her car, but he drove. "I've had enough shocks for one day", he explained. She pouted but smiled enjoying the easy banter. He really could be charming when he chose to be. They parked near the entrance to the mall and started to walk through. Walton was amazed. He hadn't expected furniture stores anymore, but the whole atmosphere of the place was changed.

Gone were the posters, replaced by china graph slogans across the windows. On many of the doors was the notice "Bring your own containers." They had kept their own selection of plastic and glass jars and bottles, carefully hoarded away in a kitchen cupboard.

Grace left him to quickly go into the drug store, returning empty handed. "No joy?" queried Walton.

She shook her head. The disappearance of tampons had embarrassed her. Walton went into the electrical store seeking batteries. There were some, but now they were just bare metal on the outside. They'd have to do. The supermarket was like a scene from an old Western. Gone were the neatly laid out shelves of garishly packed goods of all sorts. Metal bins had replaced them, each shopper filling his or her container and taking it to a pair of scales. There the assistant would weigh it and take the cash. The checkouts were gone -- redundant. Carefully, Walton poured some sugar into a jar, paid for it and placed it gently at the bottom of his plastic bag. They were precious now, hard to replace since the cut down in plastic production. One tear and it would be useless. Grace had a string bag, but that too was almost irreplaceable these days. Again, it was a fibre.

The store detective stopped him at the door, checking his jars for the assistant's personal mark that nowadays was a makeshift receipt.

Walton looked at Grace. She shrugged her shoulders. She was right, he hadn't known it would be like this.

At the butcher shop he turned down Grace's suggestion of hamburgers. She queried it and he explained that out of sheer principle he wasn't buying hamburgers any more unless he could be guaranteed they didn't emanate from Brazil. The butcher was puzzled. Walton explained how jungles were being ripped apart so that cattle could be bred for cheaper meat in America.

The butcher was shocked, but unrepentant. "These days it's all most folk can afford. There's not too many buying T bones." Walton bought bacon instead.

On the way out he went into the tobacconists. "Ready rubbed, or flake sir?" He explained he didn't smoke a pipe and opted for cigars.

"Which colour sir? Most of the ladies now seem to prefer the blue ones." Walton said he just wanted normal tobacco coloured ones. The man obliged. "You got something to put them in?" he asked. Walton shook his head. The man sold him a tin.

Grace was across the mall, gazing into a dress shop. He joined her. She was shaking her head. "Have you seen the prices?" she gasped. Walton looked, but didn't know about such things.

She explained. "They don’t have tricel anymore, nor rayon; it's either just cotton or wool almost. I guess you couldn't see me in tweed, could you?"

He was silent. The whole thing was depressing. Half the shops were empty and in those that were still open, the shopkeepers had the haunted, harassed looks of people about to go bust. No longer was it the chorus of "Have a nice day". Now it was "I'm sorry it's the shortage" or "I can't help it -- all we can get."

Grace needed some repairs done in the flat, but the shop didn’t have a proper screwdriver. The hardware store clerk apologised. It was all metal or nothing. He knew that wasn't much good for electrical work, but even the typical yellow plastic handled ones derived from a wood base. He was hoping soon for some with cloth covers, but deliveries were slower now, wholesalers cutting down deliveries to ensure that every lorry, expensive on fuel, had a full load whenever it went out.

Walton took Grace's arm and led her from the mall. "This is crazy. How can people go on like this?"

She agreed, recalling Joe's remarks last night that cities were becoming almost uninhabitable.

He'd heard Powell's piece from Kansas, and it had given him the germ of an idea, one that had followed on from the discussion in the scientific committee about social forestry. Could it be, they'd wondered, that eventually there would have to be a mass exodus West again, to repopulate the plains, to become a largely agrarian nation again? Walton last night had thought it unlikely. Now he wasn't so sure. Maybe it could happen. Maybe it would have to. Cities depended on a complex and detailed structure of supplies, transport and communication. It was all breaking down and seemed to be accelerating.

Back at the flat Grace cooked lunch while he watched an old movie on TV. Thank God they still had those, though Hollywood was really in trouble now. Film depended on celluloid and, as yet, nobody had come up with an answer. Video tape was around, but there was still enormous resistance to it, despite the almost total use of it by television.

Now, even video tape was threatened, due to the fibre content of the tape itself. More and more, TV was live, direct electronic transmission. Some people loved it with all the mistakes it showed, and they were always good for a laugh, but it somehow typified the return of the nation to a certain raw state.

And God how he missed books. Grace had kept her standard Bible, plus six, but he'd read those a dozen times now.

They'd exchanged them between themselves and their friends and had some fun along the way analysing why particular people had kept particular books.

One of Walton's assistants had been shrewd, hanging onto only the remains of his collection of soft porn. Another had kept only foreign language primers and as more and more people considered emigrating, he'd had trouble retrieving them. There was, of course, always the black market. But a hundred dollars for a book? It was crazy. Yet people were still paying. It had changed social lives too. Libraries of video cassettes were thriving, and a copy of a good film now was fetching a good price. Often people could be seen with a DVD machine in the back of their car, taking it to a friend's house so that a favourite movie could be copied. And now even music was drying up. No more could the record companies get hold of the raw materials for records, whole orchestras had their sheet music either stolen or confiscated for libraries. The amateur orchestras had been the hardest hit victims, often having to cancel concerts for lack of music to play from. The whole face of the music industry, the art form itself, was undergoing a drastic re-think, with only the genuine musicians amongst them standing any sort of chance of survival.

He brooded for a while, changed channels and caught a news bulletin. Now there was a crisis in the shipping industry.

The previous slick movement of goods was under threat because the idiotically simple device of wooden pallets had vanished. They'd been stolen wholesale, and now old time dockers were being called back into the ports to advise the younger generation of how they did things in the old days. Container ships were being converted into bulk carriers. But to that particular cloud there was at least a silver lining. No pallets and containers meant more workers in the docks, and that had to be good news. There was other good news too. The latest survey of health trends showed that people were fitter now than a year ago. They were eating less, getting more exercise and fresh air.

There was a significant saving in health care, and this had also been matched in vastly reduced bills for garbage disposal. Taking just the paper and glass out of domestic refuse -- the paper for re-cycling and the glass for re-use as containers, had slashed municipal costs by two thirds. There was even less food being thrown away. The bulletin ended with a return to Hollywood and Walton smiled cynically. A young starlet, obviously with an eye to publicity, had threatened the city fathers with a million-pound law suit if they confiscated her Jacuzzi tub. She reckoned it was the only way to keep her figure in trim, and without it, the offers of film roles would start to dry up. Since the news broke, the station reported, she had been more than over-whelmed with offers to share the tubs of others, providing, of course, they could keep her company.

Grace, just finished the washing up, saw him smiling. "Dick Walton you're a dirty old man -- what you'll need is air and exercise. Come on we're going for a walk. It will do you good," He groaned, then agreed.

In the cool of the evening they walked in the park, once again surprised by the number of people out taking exercise -- joggers by the dozen, dog walkers by the score, fathers with children -- the first time for years they'd had time to be with them properly.

Walton studied this with surprise "You know, the whole American way of life is changing. Not so long ago they'd have been pouring out along freeways, clogging them up and getting frustrated and angry with each other, Now look at it."

Grace agreed, "It certainly looks healthier, you have to say that for it." Silently she wondered how long it could last like this -- not just the new way of life for the country, but for them as a couple. Soon, she feared, he 'd be restless again. Already he'd talked of how he was most useful, buried away in a laboratory or a research station, or out amongst the remaining skeletons in the forest.

He was an odd man, she mused, flippant mostly, yet with a driving intensity that sometimes frightened her. Charming mostly, yet sometimes so forgetful, cold and hard that it hurt. It was almost as if inside him there was some devil determined to remain hidden and untouched in case it could be touched by contact with another soul. And, in those times, she couldn't reach him. She wondered if anyone could.

Idly she contrasted it with the outgoing, almost uninhibited nature of Bob Powell. full of humanity, a caring man and, if she were sensible, a much surer bet for happiness.

She'd been silent for a while, then Walton spoke quietly. "Thanks." She was puzzled.

"Just for giving me time to think. Not many women do that for a man."

They went to bed early, but Walton woke, in the early hours, his brain wide-awake. What had he been doing today? Shopping, watching TV and walking. Yet around the country there were people in trouble, worried for lack of jobs, even food, In Indonesia they were certainly worried about food, desperately hunting for fuel to heat what food they could get. Yet today he'd had the choice between steak and chops. What was he doing about it?

It was his job to try to solve it somehow. Dammit, he was even being paid for it, OK, so it had only been one day off and almost certainly he'd benefited from it. But where was his self- respect? How could he face Lestari's father? And Lestari? He remembered her love making vividly, thought much of it had seemed to pass in a blur.

He looked at Grace. One day he'd just have to make up his mind, but meanwhile there was work to do. Maybe there was some lesson from the past somewhere -- after all old men were now going back into industry to revive old skills. Maybe that was the answer, A remedy lost somewhere along the way? If only he could get to a library. A certain number of scientific books had been retained round the country, each university carefully hoarding a unique selection of knowledge. But where to start?

Still tortured by the thought he dozed back to sleep, met by a nightmare of thousands of people wandering helpless across deserts, little children clutching a mother's hand, looking up to her for re-assurance, above them in the sky a blood red sun burned down and beneath them the earth bore the scars of dried gullies where the water had abandoned the soil.

And here and there, among the crowds, evil men with wide, sharp grins offering handfuls of food that nobody could afford, and there too, somewhere, was Nani. She was crying, she'd lost Lestari. He stretched out to reach her but couldn't. He heard Lestari's cry as she searched for Nani.

He woke, sweating, the dream still with him. It was almost dawn. He lay there, relieved it was only a dream, but still shaken by its reality. Grace moved beside him, Walton wondered how he could tell her their weekend was at an end? She'd so treasured it, looked forward to it like a child. Maybe she could help him? Maybe Simon could?

He dozed again and this time it was Grace who woke him with coffee. His face still showed the troubles of the night. Her heart dipped. She knew the signs. Before he'd even said a word she knew his mind was far away -- off on some journey into a hinterland that only he knew, He smiled up at her, almost pathetic in a boyish sort of way. She knelt by the bed, cradled his head in her hands and kissed him gently on the forehead. "You don't have to tell me. Don't explain."

He tried to pretend. "Don't have to explain that there's no better place to wake up, no nicer sight first thing in the morning?" Her face stayed hidden from his over his shoulder while she sobbed a quiet tear in her mind.

And almost she prayed, a silent pleading, that he'd stay. If he went now ... He pulled her to him, she asked almost formally how he'd slept, knowing the answer, sensing instinctively the answers to come. He pulled himself up on the bed, looked her in the eyes and held her hands. "You know don't you? I have to work. I have to try"

She considered the options. She could feign surprise, but he already knew she was going to let him go. She could protest, but since they both knew he was going, what was the point? It would be a charade. She could just accept it, but then there would be no demonstration that she cared.

Instead she just nodded. "I'm trying to understand. And I'm hoping too ..."

"Hoping?" he asked.

"Hoping that at the end of the day you'll want to be back here." Now, and they both knew it, he had to dodge, to evade the issue. Either of them could have written the script.

He said the words "You know I'II want to be back, and with any luck I will be. But ..." Silently, and just to themselves, they each finished the sentence.

*"But if I get involved, if the work takes me over, then you'll be forgotten and maybe later, someday, we can try to repair the breach. We can pretend that this weekend, this treasured promise that we'd put aside, was shattered only by an emergency, an outside force, that had nothing to do with us."*

Stupidly, Walton felt like weeping. What difference would a day make? And was work that important? Or was it work?

Wasn't it just that one more day of being like this and his defenses would be breached, his loneliness exposed, that he'd end up saying 'Grace, I love you', 'Care for me, protect me, mother and pander to me, be like all the other wives are to all the other guys? What did he have to be ashamed of?

But he knew he'd never accept it, not 'til it was too late anyway. Yet if he were to collapse -- his mind's thoughts, not his -- wouldn't that same question still be there: wouldn't it all be the same as it had always been with other women?

Wouldn't that hidden devil of selfishness, laziness and secret inner self sneak its way back to wreck it all, slowly sour it all over the endless years to come? Almost detachedly, he'd see himself destroy someone he loved. How could people have the courage to let others that far into their minds, didn't they need privacy for their souls?

The spoken words still echoed round the room. *'You know I'll want to be back, and with any luck I will be. But...'*

They'd come only mini-seconds ago, yet Grace was already facing her truths. Of course, she loved him -- totally. He'd been her only lover and that alone made him important. So often she'd glimpsed through the partly opened door of his soul, seen the fullness of what was there. But that door always closed so fast -- Dick didn't want anyone to go through it. He was scared of it -- of himself.

She'd gone on loving him, hoping one day he'd acknowledge the bargain love was about. Right now, she still had the strength, But years of this? What would there be of her left? What would she have to offer? It was a sweet irony. Had he just taken what she offered, ravished it crudely and discarded it, she could have borne it more easily. But he hadn't. He had, in simple terms, kept her dangling and that almost made her bitter. If only he'd been a plain, straight bastard it would have been so much easier. She felt like the orator at a funeral, the death of a love affair -- the bitter end of sweet.

The breakfast was poignancy; eggs and toast. Neither of them acknowledged the parting. Walton went to the library. Grace had a million things to do -- all of them alone.

Just before lunchtime he called her from the library. He thought he might be on to something. But it meant being delayed. Grace gave him the excuse -- some girlfriends had phoned her, it would be all girl talk 'til late and wasn't it time he had a look at his own flat? And an early night, up bright and sharp the next morning, wouldn't that make sense just for once?

He put in the formal protcstations and in about five minutes it was all over. Grace wondered what she'd have said if the girls had come around. She could, hardly explain it to herself.

***

## CHAPTER FOURTEEN

Lestari went just before dawn, leaving Walton to sleep off his jet lag in the guest cottage. He had to be fresh, she knew, for a meeting later with her father.

Walton's return had surprised her, for although he'd been warm on the daily phone call he'd also sounded pre-occupied. Again, at the back of her mind she wondered if there was another woman. He had, she knew, been working almost non-stop, driven on as if America's guilt was his alone. She'd tried to explain it to her father, but he'd merely given her one of his wise old nods, He knew her judgment wasn't entirely detached, but he said nothing. She had needed a man and now she had one -- she'd chosen him, and it was up to her.

As Walton entered the old man's office he looked anxious, perhaps slightly nervous, the old man mused. As always, Walton wore that indefinable feeling of guilt. Secretly the old man smiled to himself, wondering if the guilt related to the plight of the country or Walton's now quite obvious relationship with Lestari. Either way it was irrelevant. The American was here to help. And he had been useful.

Together they reviewed events since their last meeting. Walton told him about American welfare payments were running out, the concern over the increased silting of dams and the way gasoline was now restricted, both by price and import restrictions; and that there were White House thoughts of trying an experiment in community farming with people moving from the cities to the countryside.

"It seems" said the old man, "there are signs of a change from materialistic attitudes and consumerism toward neighbourliness and religion." Walton agreed it was a trend though he doubted that such ingrained attitudes could disappear overnight.

He explained "There are still people bemoaning they don't have charcoal for their barbecues? 'They have a million petty complaints. Sometimes it makes me despair."

The old man smiled "It will take time. Would you expect people to become accustomed to poverty in a day?"

"But, your problems are so much bigger" said Walton. "Nobody in America seems to understand that."

The old man agreed that might be so for the people, but the government was exempt from the charge. Walton queried it. The Vice President went on "I believe, after our last talk, you told Mr. Simon of our problems. Am I right?" Walton nodded agreement. The old man continued.

"By the wave of a magic wand we suddenly had all the help we could have asked for from such people as the UN, its agencies and the World Bank. I think Mr. Lean must have helped there."

They weren't the only ones to have offered help. The old man sketched in the offers -- and the strings attached to them -- that had come from the Russians and the Chinese, He'd had to accept them. He explained "A lot of our people are now in a poor state of health."

"Can't you get them drugs? You said there was help from the World Health people." Walton frowned.

"It will take time for them to get them. You see for centuries my people have made their own remedies from the herbs, the plants, the barks of trees. If they had cascados -- a disease of the skin -- they knew which plant would cure them. They'd learned it from their ancestors. It was the trusted way. The foreign doctors tried of course, but if they gave them a pill the doctors would go back a week later and find that person wearing it round his neck. It does take time, but meanwhile they suffer, cultures of all ages have their different problems. Do you appreciate how primitive some of our communities are?"

Walton didn't. The old man looked wistful. "You are not alone. One man from the UN had a wonderful plan to instruct the people. He thought radio broadcasts and pamphlets were the answer. We had to explain that some of our people still lived in caves, some even in the tree tops and had never talked to anyone outside their own tribe and in their own language."

Walton had to ask "People living in trees? What happened to them?"

The old man didn't know. "They were always nomads and we're still trying to find them. They seem to have hidden somewhere. The only reliable report we had was that they all got into their boats, sailed away and said they were going to look for the leaves."

Walton looked troubled.

The old man re-assured him, "I think, maybe, they will be amongst those who survive. Our problem is the great mass of the people who are hungry. You sec, the first rice crop was a total failure. The rain that came flooded down and swamped the plantations. The water was too deep for the rice. In the old days it would not have been so bad, but this rice was a new high yielding variety that had short stems. We had been told we would get three crops a year instead of one. Now we have lost a whole harvest." He sighed, "It seems the world is full of ironies."

Walton still looked puzzled. The old man continued "When the floods stopped, we then had far less rain than usual. Now, we are told the river levels are dropping and, in a few months, if there is no relief, the salt water will start to encroach further up the estuaries and that means much more infertile land. I think you will possibly have the same problem in America?" Walton agreed there was a theory on those lines.

For a while they both sat in silence, digesting the scale of the crisis. There was little to be said. As if recognising that, the old man rang a small bell by his side; there was a moment of incomprehensible chatter and moments later Lestari and Nani appeared, closely followed by a servant with a tea trolley. The atmosphere changed instantly to become a family party. And, thought Walton, for once he didn't feel too badly at all being part of a family.

This time Nani had a present for him, carefully wrapped in a piece of coloured muslin. She watched, wide-eyed, as he went through a pantomime of what on earth it could be. She jumped up and down in excitement as Walton very slowly unwrapped it. It was a packet of cigarettes; they were Russian. Lestari explained, that day after day, delegations arrived in the capital for audiences with either the President or her father and, knowing of the shortage, they brought such gifts with them.

She smiled "Wasn't it the English who used to give the natives beads? You should now regard yourself as a second hand native."

"Smoke one. Smoke one now" insisted Nani. Walton put up one finger and explained he would treasure them for later. He could hardly explain the occasion.

It was the one thing he missed more than anything; his cigarette after lovemaking. Lestari smiled to herself, knowing what he was thinking.

In bed that night, lying, relaxed now, in the crook of his arm she teased him "Do you think the Russian Ambassador would approve of his gift being used in such a way?" Walton laughed out loud, drew again on the cigarette and said "Here's to the Kremlin."

Casually, she asked if this time he would be here long. Walton considered the question gravely and pulled her leg slightly. "Leave a place with free cigarettes? How could I? But seriously.....? Seriously, I have this odd feeling that somewhere there is a very simple solution, almost staring me in the face perhaps. And here I can concentrate so much better on work. I don't have to go to meetings for instance. And all the time in America there are reporters hanging around. You daren't share any knowledge in case it leaks out, particularly in Washington. Here it is much more secure."

Lestari turned to him. "Are there no other reasons for being here?" Walton kept it light hearted but held her warmly. "There are some great reasons for smoking cigarettes."

They slept then and, again before dawn, she slipped across the garden to the house, observed this time by her sleepless, worrying father. At least now she is happy he mused and whispered to himself, *'I hope it lasts'*.

***

It was Joe Simon's call that got Walton out of bed. There was a meeting shortly of the Scientific committee and it was getting worrying reports of soil erosion. How bad could it be?

Walton, still half asleep, groped for a cigarette. Simon recognised the sound. "You have cigarettes?" Walton told him the story. Simon laughed outright, "Can I tell the President? That would really appeal to him. Maybe you could bring him back a packet?"

Walton promised to remember though it struck him that if the most powerful man in the country couldn't get himself a smoke then the world really was in a state,

He'd had time to wake up now. Erosion...he flicked through his brain. "From memory I think the best comparison is with the Philippines. They cut down their trees because they needed the money and when I last heard about 40 odd tons per hectare were going down one of their main rivers."

"That a lot?" asked Joe. Walton told him 'I think the figure for the Mississippi was around half a ton. The problem is -- and it's happened all over -- is that reservoirs and hydro-electric schemes get fouled up very quickly. Again, from memory, I think it's Columbia, or somewhere in South America, where they have to ration the water, and the electricity, because the water runs straight off the hills. And I read somewhere too that the Panama Canal is filling up pretty fast too. That's bound to happen when you don't have tree roots acting like a sponge to hold the water. Oh yes -- and there's another problem too."

Joe groaned "Give it to me -- I might as well have it all."

Walton went on "The water is pretty dirty, and so you get more disease. It's a great way of spreading malaria." Joe groaned again. "Don't tell me -- you treat that with quinine and that comes from trees in Indonesia and they don't have those trees anymore,"

Joe sighed "It could be a real fun session today."

Walton interrupted "While you're there you might get someone to explain about estuaries salting up when the river-level drops."

Joe almost shouted "That's enough."

Walton asked him what was happening in Washington, Joe said he thought the President liked the idea of the farm sharing project. Oh, and they were going to try to import some old-fashioned hill sheep, plants some grass on the high land and see if that helped. It could bind the land, ease the erosion problem and also provide some wool for the textile trade. That industry had suffered through lack of Tricel and it needed a boost. It helped with jobs, But the farm sharing idea was the important one.

Simon couldn't resist a joke at Lean's expense. He told Walton "I think John has it all planned. Ten-gallon hat, covered wagon from a museum and a great cry of *'Wagons Roll'*. I guess he can hardly wait." Walton burst into laughter. Simon certainly knew his man.

They said goodbye and Walton stepped into the shower. He was almost used to the dirty brown water by now. It smelled of chlorine, but at least it was hygienic. Soon, he guessed, they'd have the same problem in America and boy wouldn't that cause a fuss.

But then again someone would invent some new cosmetic and convince people they couldn't possibly shower without it. Someone, somewhere, would be bound to make a fast buck; it was that sort of place.

And here they just left their homes in the trees, got into their boats and sailed off in search of the leaves. It really was two different worlds.

***

In Washington Simon went to the Scientific Committee meeting. Grace, as its new secretary, was already there. Simon mentioned he'd just spoken to Dick and told her he sounded fine. She didn't doubt it.

The previous night she and Powell had dropped in on an embassy reception where one of the diplomats had just returned from Indonesia. He'd met Walton there and teased him about his ability to hire stunning looking girls to help him. Grace in Washington; Lestari in Jakarta. He'd meant it as a compliment, but for Grace it was confirmation of her suspicions.

It had now convinced her that she and Dick were washed up for good. Joe Simon broke into her thoughts, "Hey young lady -- you're dreaming. There's work to do. I've been hearing about this erosion problem. I never knew it could be that bad, I just don't know what we can do about it."

The committee discussed it most of the morning. The only possible solution seemed to be a massive operation to dredge the dams and the rivers and, if possible, return the silt to its original home. Was it feasible, the committee wondered, to replace the topsoil from the prairies with silt from the estuaries?

Also, they concluded, they needed a lot more irrigation if there were to be small farm units, developing a greater variety of crops. For Simon that settled the issue, the dams and rivers would have to be dredged.

He wondered though about the salt coming back up the estuaries. Could that affect the plan? Few people had direct experience but those with UN contacts would check it out and report back. Meanwhile Joe would report to Lean.

***

The President was not in the best of tempers. Joe lightened the mood with the tale of Walton and the Russian cigarettes and hoped he could return with some.

Lean was more cheerful then. Cigars were OK, but he'd been a 40 a day man and they just weren't the same. How was he supposed to concentrate on solving the most urgent crisis in the history of mankind if he couldn't have a smoke? He'd tried for them all over but was terrified his attempts would leak out.

Joe summarised the committee discussion, outlining the problems and the proposed plan of action. The longer Joe went on the more cheerful Lean seemed to get. What attracted him was the idea of vast dredging and digging operations. That was a big job. It appealed to his sense of the dramatic. He had, and admitted it, a love of the big occasion; and the more Barnum and Bailey it was, the better. It was the whole thinking behind the Alaskan bombing and now he could do it all over again. In seconds Lean could see it all -- him in a hard hat sitting perhaps in a huge digger, emblazoned perhaps with the Stars and Stripes. Maybe he'd drop a flag and send off a long line of them to work. He'd check with the Press office to see which would make the best pictures. Joe looked despairing, but Lean was unabashed.

Politicians at work rarely made good pictures, so they had to grab every chance they could get. There were barely any newspapers left, but they could time it to get the best of all the TV newscasts. Wasn't there some old British phrase from the Second World War. That guy Churchill had known about good pictures. He'd look it up.

"It was dig for victory" said Joe, resigned to it now, "it was about growing your own vegetables in the back yard."

"It's near enough" said Lean " and it has a positive tone to it, and don't forget" he said "Churchill won."

Joe mentally shrugged his shoulders. He'd come in to talk about electricity and water shortages and all Lean had talked about so far was cigarettes and hard hats. He hoped the President knew what he was doing.

Joe managed to complete his review, reporting that the arrangements were now going through for the farm sharing scheme

Lean took it all in, though with no great show of enthusiasm. He slightly queried the rent a farm scheme. Lean explained "Some of the politicians out there have been getting some anxious noises from the farmers. They're still worried about who'll be going out there. Are you sure that won't cause trouble?"

Joe knew there were dangers. "But we don't have any options and we do have an alternative or extra scheme going as well."

"What's that?" asked Lean, Joe told him the idea of putting more sheep on the hills, giving them rough, soil binding grass to graze on.

Lean was dubious "That doesn't give many people jobs Joe."

Simon countered "No but it does provide wool and that keeps the textile industry going so it has an effect down the line."

Lean took the point. "How soon can we get all this going?"
Joe wasn't sure but agreed it should be as fast as possible. In the current crisis they might get a law through in a week.

"Any particular reason for the rush?" he asked.

Lean nodded "You see the opinion polls lately? We have to be seen to be doing something now. We've had a lot of stick lately for all sorts of things and don't forget before too long we'll be into mid-term elections. Now those could be real bad."

Lean rose from the desk as if business was now over. He went to his cupboard and got a bottle. One before you go?"

For once Joe said "Yes,"

Lean raised his eyebrows. "This getting to you?"

"Sort of" Joe confessed. "Everywhere you look there's a problem. I think I've almost qualified for a science degree lately. You know I failed that in high school? Nearly blew it up once." Joe took his drink.

Lean looked at him carefully. "Joe I know I get impatient sometimes and I know you think I'm crazy tearing around with TV cameras and the rest, but if people out there think we're just sitting on our butts giving out bad news while they're suffering it doesn't do anybody any good. I never forget when I was a kid the President came to town and that was a big, big day. I guess I never forgot it and I've got a feeling that's not changed too much. People still like to see the President out doing the job,"

He poured another drink, offered Joe another one and smiled at the refusal. What I'm also trying to say is that I couldn't do that job if I didn't have someone back here doing the real one, running the machine, I just wanted you to know I don't take you for granted."

Joe confessed "I guess I'd look silly in a hard-hat anyway. And, thanks."

"Hey" said Lean I nearly forget the good news."
Joe said he'd almost forgotten it existed.

Lean beamed "I leaned a little on the South Koreans. I guessed they still owed us a favour. Did you know they have a national tree day?

Anyway, they're shipping over a whole load of seedlings for nothing. Imagine --us taking handouts from them. Over there they give 'em away to kids. I thought we could do the same. I wonder if they'd grow out here on the lawn?"

'Oh God' thought Joe 'he's off again. Is there a spade in the house?' Out loud he said "It's a good place to start. But John, just one favour please?"

"Sure Joe -- What's that?"

"Can we just make really sure someone takes care of it. It would be a real bad place for it to die."

***

At the radio station's morning conference, the news editor, was cursing the heat. It was more like August than June and it had been like that for days. People talked of nothing else and in his eyes that meant it had to feature in the day's coverage. Weather stories were good -- and easy.

"Jim?" The crime man lifted his head, muttered "Yeah, I know -- hot days, more crime. Downtown ferment – just give me about a minute or so."

The editor glared at him, "But no panic stuff -- we don't need people saying we started something."

Jim slowly put down his coffee. "Panic's too much trouble. But I'll tell you what's going on. We got ghettos here right? Full of young blacks with no money and lots of energy. They hear things are better out West. They got food and things and maybe there's a place to grow stuff. And they've all got this roots thing about going back to the land. They saw it on TV. So, they got motive. So, the crime of the week is campers. Steal one, drive west into the sunset, country and western music, head full of dope, maybe your lady -- or someone else's -- and away into the wild blue yonder. It's keeping the cops real busy."

"At last" the editor sighed "something I understand. Get it to me."

"Now what else?" He needed someone to beat over the head. "Casey. You're the science man. Why's it so damned hot?" Casey, still young enough to be flustered looked lost, "I tell you why" said the editor "It's something to do with the weather. Get the weather service -- I want half a minute."

"Bob -- what's happening on Capitol Hill?"

Powell grinned back at him. "Mostly they talk about the weather. The editor didn't see the joke.

Powell hurried on "I'm not promising, but there was something I heard last night just could tie in with the weather."

The editor nodded. He could trust Powell. The conference broke up and he got the number of his science magazine contact from Casey.

Powell admitted it sounded crazy, but at a party last night he'd stood next to three scientists -- all of them on Joe Simon's committee -- and they'd talked about greenhouses. When he'd tried to join in they'd split up and shut up. Did it make any sense?

The writer knew exactly what the story was. If his magazine had the space they'd have been doing the story. "If you can't do it you might as well let me -- and we'll credit both you and the magazine. OK?" The man agreed, "So what's the story? asked Powell, switching on his recorder.

Slowly, as if he was telling a ten-year-old, the science writer went through the details or the greenhouse effect, increasing carbon dioxide providing a blanket, the effect of which was to warm up the earth, maybe eventually melt the Antarctic ice-caps.

Powell stopped him, "Look that's not new. People have been talking and writing about that for years. What's new about it? Does it explain the weather we're having now?

The science man wasn't certain. "It's difficult to say, but it could be. One thing's for sure, if the temperatures do go up it changes the pattern of the world's agriculture. For instance, you'd get much warmer and drier weather out on the corn belt, although it would be more difficult to grow stuff there without a lot more irrigation than they use now. But you know what really worries me?"

Powell couldn't imagine. The writer went on "Even if the trees hadn't gone and the government had woken up to the risk it would have probably taken them round 20 years to change over to different fuels; away from oil to solar energy. You're talking about never using gasoline again in cars, not using coal, the whole thing. Now, say they worked it out that the crucial point for that change was the year 2030....."

Powell got the point this time. "That means we should have stopped using fossil fuels in 2010. Oh boy...."

The science man reminded him "Don’t forget that was if we'd kept all our trees. The only question we're arguing about now is how soon and how drastic it'll get."

Powell put down the phone and quietly took the editor aside to explain what he'd got, "Shit" whistled the editor. "Some happy weather story."

The editor had a nasty second thought about the Government’s new censorship rules. Powell was aghast. 'They couldn't stop it could they? By what right?

The editor raised his hand. "Bob, I couldn't agree more, but I'm a born pessimist that's all and it smells like trouble. Don't forget; the rules now say if we think its sensitive we have to check."

"So, we didn't think it was sensitive, a lot of it's been in magazines already," Powell argued weakly.

The editor made his decision. "“We give it to Casey as a background piece with leading authorities and what do you know, this comes up in the middle, that way we could argue it was just incidental."

## CHAPTER FIFTEEN

In Jakarta Walton had to brief the old man. He agreed it was serious but pointed out that there was now far less carbon dioxide going into the air from cars and factories.

The old man acknowledged that fact. "I think the American phrase is 'not before time'. Is that so?"

"Yes" said Walton, "but the governments can't really say it's taken them by surprise. There's been so many reports over the years they had no excuse for not doing anything about it."

"Except it didn't seem the best way, at the time, of getting votes did it?" chided the Vice President.

Walton gulped slightly, hoping he wouldn't offend the old man. "No more so than ensuring the multi nationals obeyed your forestry laws more stringently. Some money spent there might have helped a little as well,"

The old man sighed and spread his hands. "It wasn't easy for us either. The rich people wanted our wood. We were poor, and it was our only way of getting food. We had hoped that one day we could slow it all down, but the pressure for food was always there. It still is..."

Walton left it there. Hc thought ahead. The weekend was coming up and while he wasn't taking any time off he was at least combining business with pleasure. Lestari had nagged him about not seeing the rest of the islands. It was true they didn't have the beauty of former times, but it would afford them time and place where they could be truly alone.

But first Walton had to ring the office in Washington and tell them he'd be out of touch for a day or so -- a field trip he called it.

Grace had taken the call -- just as any secretary takes a call from her boss.

Walton put down the phone and went back to his microscope. He could see things clearly with that. Women he couldn't. The way Grace had spoken you'd have thought she only ever been his secretary. Maybe there was someone else? He dismissed the thought. No, not Grace. She had a peculiar sort of loyalty. One of the things he liked about her. He could sort it out when he went back next time. It might take a while, but he knew how to charm her. He'd always known how to do that.

Grace had the drinks poured when Powell arrived. They got over the awkward moment where he dithered over whether to kiss her or not. She was more relaxed now, some of the strain had gone. Powell wondered what had caused it. He doubted it was work. It had to be Walton. He made an excuse to go to the john.

Just as he'd hoped, there was no razor there anymore. The shaving foam, too, had gone. Flushing the loo, he used the sound to cover up a quick inspection of the laundry basket. No sign of any male garments. He felt guilty, but relieved too. Maybe this time Walton really had moved out.

He re-joined her in the living room. "I'm sorry" he lied "I barely had time to wash when I got home." She was standing close to him. He ached to hold her. He felt about sixteen -- totally unsure of what to do. She sensed it -- and broke the mood.

"I'm starving" she said, "take me to dinner right now."

Powell aped a servant's wave and ushered her to the door.

***

Throughout the dinner they maintained the banter, jokes overlaying messages they couldn't give each other out loud. He knew she felt happier that way. They talked too about the White House, a consuming interest for both of them. As usual his ear for the malicious titbits of gossip kept her amused and, along the way, amongst the bonhomie, she told him Walton had returned to Indonesia.

She didn't know when he'd be back, but he'd be in touch. 'Bosses always keep in touch with their secretaries' was how she put it.

Powell got the message, but still wondered what had happened.

She looked now like a girl who'd made a decision and was glad of it. It was only his instinct, but he was sure Walton was now out of her life... at least as far as she was concerned.

He wondered how quickly she would recover. Sometimes it took a while. So maybe he shouldn't push it too hard tonight. He didn't know. Perhaps she wanted to plunge straight into something else as a way to forget.

She teased him now -- about looking thoughtful. He put on his mock serious air. “I was just thinking I should tell you the things you have do to stay on my list.”

“That” she said “is a very chauvinistic attitude. I think maybe you should hear what keeps a man on my list?”

He considered the question for a moment, staring deep into his brandy glass. “You think this is the right setting for such an important conference? I move for an adjournment. My place or yours? The decision is up to you.”

“Can I raise a point of order?” she asked.

“I guess so” said Powell.

“Nobody agreed yet to the adjournment” she pointed out.

Powell looked mock solemn again. “From the chair I have to advise you it’s in our joint interests.”

“How so?”

He grinned wickedly. “There might be some practical questions and I’d like to be able to come back here someday.”

She giggled outright. “Agreed” she said, adding “my place -- on account of your disgusting coffee. You failed that test already.”

They left and drove through the empty cavern of the night streets to her flat. She made the coffee, added a brandy and joined him on the sofa.

He joked “I always knew you were after my body.” She put her hand on his stomach. “It feels OK I guess -- for a man of your age.”

He winced. “But age brings wisdom you know. It isn’t all bad.”

She was somehow closer now. “And what does your wise old head tell you?”

He turned to her -- serious now. “At times like this it’s not my head that talks to me. Maybe it would be better if it did.”

She understood but stayed close to him still. He knew now what to do. But he still had to carry through with the words. “At times like this my head will tell me all the things it should. Does yours do that too?”

She nodded, then added “It does. And then I get to wondering how hard I should listen. Maybe even whether I want to listen.”

He felt the same. But he had to explain. “Grace I am older than you. Maybe ten years or so and in that time my head hasn’t always been the best judge. Sometimes it feels as if it just isn’t there at all.”

Her hand was still on his stomach. He covered it gently, then followed her arm, stroking it gently as he went. She sighed gently and he moved her forward, slipping his other arm behind her.

She eased forward and, in doing so, her hand slipped down. "Stay there" he whispered. His hand slipped easily into her blouse, seeking her breast. As he found it her hand gently began to stroke him. He was half panting now. She moaned and pulled him onto her, his tongue seeking hers as they kissed.

They were desperate now, breath coming short, minds blurred, feeling only each other's longing. "Bob, not here; let's go to bed."

Gently she led him along to the bedroom, softly now and then looking back at him reassuringly. She stood by the bed, raised her arms and clung to him. They fumbled their clothes off, dazed still by the suddenness of it all, neither quite able to catch their breath.

She slid onto the bed and at last he saw her glorious, wanting, nakedness. He stopped then and raised her head to him. He waited until she opened her eyes, drew back his head so that he could see all her face. Almost with reverence he bent to kiss her and as he leaned down her arms went around him. And now, she thought, I'm purged. This is a new man, a new way, gentle and caring and oh, so loving.

Different people later, they held to each other. Spent and speechless, body had spoken to body and would do so again.

She sighed "Oh Bob, you just didn't know."

He was still reeling, staggered by the impact on his brain. And then they slept. They woke, returning to their self-cast roles of banter. She slipped her hand down to him, felt it stiffen again and whispered, "Did I pass the practical?"

"Madam" he said, "I think you set the exam."

She gentled into him. "Just hold me Bob. Just hold me tight."

He did and felt her glow. "You know," he confessed "I have the strangest feeling. I can't explain it, but somehow it's almost like the start of everything. It's never been quite like that before. Do you think we'll get used to it?"

She eased back over slightly. "I was hoping somehow you'd have all the answers -- you and your wise old head. Me, I just don't know. I just feel complete. And safe. And warm. And that's the first time… It really is."

***

Though still in something of a daze, Powell left and drove to his flat to change his clothes on the way to the office.

He was late. The editor was itchy. It was a slow day. One of the engineers had told the editor Powell had an important dinner last night.

"Is nothing secret round here?" asked Powell.

"Did it get you a story?" asked the editor.

Powell put it obliquely. "It was a frank and friendly exchange of views. Isn't that the way they put it? I think the answer is I have no comment at this stage."

The editor growled. "OK smarty pants. Just as long as you ain't charging it."

Powell beamed "I don't do things like that."

This time the editor just grunted -- "You did before."

***

The Presidential circus, all three rings of it, was on the road again. Lean, complete with hard hat, was doing his Churchill imitation. V signs, "Dig for Victory", the whole bit. And all that before he even got into the aircraft so there'd be something for the lunchtime bulletins. Powell sighed. It was alright for TV; they could just carry the pictures. He needed a story. And there wasn't one -- at least not one he could keep for himself. He was also galled because Casey got the credit for the world weather story, not Powell, so he had nothing to boast about. He accepted a bourbon from a White House press aide and felt more cynical with every sip. This wasn't a story -- it was pure news management and propaganda. He felt ashamed to be part of it.

The plane landed in some anonymous piece of mid-west desert; air conditioners kept reality at bay as the cars took the reporters to the site. A line of huge dredgers stood ready to perform. The press aides fussed over the TV crews. Lean, time and again, climbed in and out of the cabs for their convenience. He drove one off, hanging out of the cab to wave. He returned, got out and dropped a huge Stars and Stripes flag to send a convoy of them into action, lined abreast. He perched on a mudguard, let the sand dribble through his fingers and gave interviews, six in a row getting the Christian name of the interviewer right every time. Grudgingly Powell had to admire the man. In this he was expert.

Lean made all his announcements. There was the great trek west -- Operation Pioneer Spirit they called it now. On July 4th there would be Korean seedlings for all the school children to plant. And, he revealed, there would be a ten-million-dollar reward for a solution to PWN.

As reporters scuttled to and fro, Powell wandered to the marquee used by the digger drivers. “Is all this going to work?" he asked one. The man was unsure. “Look at it this way, you work all day and you get out the silt, then lorrics come and take it away and at this end you have machinery to spread it back on the hills where it first came from. Then it rains and you do the whole damn thing all again. It all takes a lot of gas you know. And we don’t come cheap. It’s OK when it comes to digging ditches and irrigating things, but this way I’m not so sure. Anyways there’s one good thing about it.”

“What’s that?” asked Powell.

“We sure aren’t going to run out of a job the way the dirt comes down the rivers now.”

Powell felt better for that -- at least he could now give the press aide a hard time if nothing else.

***

The cars purred them back to the air base and they took off again, this time heading North East, for Detroit. There the reporters filed obediently along lines turning out solar heating panels.

They were now in full production, and panic stories about electricity shortages were boosting sales. The President performed with them, tucking a couple under his arm for use in the White House. And he had more good news. With the dollar weak they were grabbing the export market. It was good old American know-how in action.

Powell did his duty recording of the speech, but once again felt he was being used. He detached himself from the crowd and spoke to one of the men on the line. "Seems a good idea" he volunteered.

"Maybe" said the man "but it don't beat making cars."

“How’s that?” asked Powell. The man looked at him as though he was stupid.

"Think about it boy -- what happens with cars?"

"They go along roads and get places” said Powell. The man nodded. "Then what?" Powell was stumped.

"Then someone brings out a new one" said the worker. "The old one wears out, so you get another one; it means repeat orders. These don't. Can't see why the President didn't think of that."

***

An hour later they landed back in Washington -- just in time for the evening newscast. Powell was tempted to use the dissident comments of the workers he'd talked to. At least he'd feel like a reporter that way. Then, in the booth, he heard the crime reporter's story, a follow up on the young blacks stealing campers.

Powell suddenly realised they'd be listening. How could he tell them they'd be going on a wild goose chase? Instead he just related the events of the day.

As he walked back in the newsroom the editor looked up. “For you that was pretty dull. Where's the old Powell bite? You getting soft?”

Powell explained. The editor grunted. "I see what you mean. They've been ripped off enough. No need for us to take away their hope as well."

***

Walton woke and felt Lestari beside him, yet it was light. By dawn she'd usually left the cottage and gone back to the main house. A minor panic invaded his brain, he half rose and looked at her. Her long black hair spilled over the single white sheet that covered them. But there was a mosquito net. Where the hell were they?

He sat up, then remembered. A desert coral island, dawn, a naked native girl -- and she'd brought some cigarettes.

There were just sometimes in life when everything turned up trumps. He wondered if she'd wake soon and get him a coffee. She gave half a sigh, turned slightly and murmured. He grunted back. "That means you want me to get up and make some coffee? He grunted again. She slipped from the bed, knelt down at its side and kissed him. "Then we'll go for an early morning swim."

He groaned "I'd prefer coffee, a quick trip to the john and coming back to bed with you."

She stood up. "You can be very unromantic at times."

He nodded. "We can swim any time."

She returned with the coffee and found him at the door. "This is quite a place "he remarked. She explained it was reserved in the old Dutch colonial days for VIPs.

They slipped back into bed and Walton drank his coffee. He was coming around now. She snuggled against him. "I think you had a good idea just now -- you know about making love?"

Walton surprised her. “I’m not so sure it was." She looked baffled, then he grinned. "Do you know I've always had this fantasy about running naked down a white beach, plunging into the sea, swimming, then making love to a native girl on the beach."

Lestari smiled "I'll race you."

***

Afterwards, stretched out on his back on the beach Walton turned his head slightly. "You realise, don't you, that you give me problems?"

“I don't mean to" she replied,

“But you do" he said, "You know I’m running out of fantasies. That could give a guy real problems."

She thought for a moment, then playfully patted his stomach.

"You stay here -- I’ll get breakfast," Minutes later, it seemed, he heard a whistle. He turned.

It was fantasy time, again. She was still naked, but bearing a silver tray of cereals, toast and coffee, "A silver tray?"

She smiled, putting it down. “I told you this place was for VIPs” She explained “This whole area used to belong to the Dutch, and they believed in treating their visitors with the utmost courtesy.”

Walton believed it.

***

Eventually, they dressed and walked round the island, Walton taking his binoculars with him. One end of the island rose gradually towards a small volcanic outcrop. They climbed steadily; lizards made startled scurries into the pink rocks. The sun was high now, just past overhead, taking away the glare. Walton scanned the horizon. Half way along, he stopped suddenly. He handed the glasses to Lestari. "We have company."

Carefully she studied the small boat as it sailed across her vision. She overtook it, looking for its destination. She stopped, refocused, then focused again.

Walton teased her. "Do you need glasses?"

He looked at her again. She was almost pale. She handed him the glasses -- silently.

"What’s wrong?” he asked.

She said "Look at the boat, then go left and refocus right on the horizon. What do you see?"

Walton did as he was told. Slowly he put down the glasses. "Now I know I'm dreaming." He looked at Lestari "It can't be true."

He took the glasses again and studied the view for a long time. "It can't be true of course. I mean trees -- with leaves. Leaves in this area?"

He gestured quickly. "Who's the man in the boat? Have a look." She focussed on the small boat, long and loaded, carrying only one man.

She shook her head. "I can't be sure."

"But what would your guess be?" Walton asked, excited now. Lestari looked at him, "I think we are hoping rather than guessing aren't we? It's one of those tree dwellers and he's found an island with leaves."

Walton almost snatched the glasses from her. "If it is and if those trees are alive as they look then in some way they've either escaped both the acid and the PWN or...."

Lestari finished the thought. "Or they've found a way of resisting it."

Walton carried on "And if they've resisted it we should be able to find out why," And having found out, she thought to herself, you'll go away,

They debated whether they could catch up the man in their own boat, then realised they couldn't. Walton raised the glasses to the horizon, as if mesmerised by the sight of the trees, Slowly and carefully he scanned the outlines of the other small islands.

All were like theirs, just volcanic outcrops, with spiky skeletons of trees poking up from the lush undergrowth. He was changed now, urgent and restless, all the languor of the morning gone. He was a scientist again, his brain whirling. They reached the hut again and Lestari began to prepare lunch.

The plane was due back an hour before sunset. Walton, impatient now, checked the fuel in the outboard motor only to find there was not enough to get them to the island and back in safety. His plan had been to go there, leave a note in the hut for the pilot and let him collect them there.

Lestari warned him against that, some of the tribes were rumoured to have turned hostile to white men since their trees had gone. They didn't understand much, but they did know it was the Americans who'd robbed them of their trees -- just as they had been doing for the last twenty years.

Walton could hardly wait for the plane to arrive. His brain was buzzing. Could the acid and the PWN have missed the island? It seemed unlikely, it had hit all the others in the area. No, the answer had to be there was something on the island that resisted the PWN. The acid must have hit the trees, but their leaves had grown again. That could only have happened if they were healthy. So, what was the answer?

The plane landed at five. They were all packed and ready -- and raced to it.

"Tell him what to do -- and warn him that he's not to report it when he gets back" ordered Dick. Lestari did as she was told, unused to the sudden steel in Walton's voice.

In minutes they were over the island with the trees. "Put her down" ordered Walton. "I must have a look at it." Lestari was still anxious about hostile natives. The pilot had a gun and they took it with them. As they neared the shore Walton leapt from the inflatable boat and raced up the beach. The trees were untouched, though they showed signs of having sprouted new leaves. Old ones littered the ground, He stood glaring around him, trying to see what was different.

Lestari wandered about ten yards away, then twenty, the pilot following her to stand guard. He knew the natives still used bows and arrows and they gave no warning, Lestari smiled with pleasure and bent to pick a rare treasure – a marigold.

Walton stamped around the jungle edge, examining leaves, ripping them from shrubs, then, more carefully, shaving pieces from the trunk of a tree and slipping them into his bag.

Lestari heard him yell. "They were here; they were. The PWN came and they died. Something killed them."

In the gathering dusk, Lestari could see him frantically shovelling handfuls of earth into his bag, ripping leaves from the trees and cutting away their bark.

She finished picking her flowers and went to him.

The pilot was getting anxious, He had to have the bare minimum of light for take-off. She almost had to drag Walton away. He was mumbling incoherently to himself. Desperately, as they circled after take-off, he glued himself to the window, anxiously trying to see if other islands had trees with leaves. He was sure they didn't. When it was too dark to see any more he found a chart and pinpointed the island.

He told Lestari "Tell the pilot to land at the nearest big town. We can stay there tonight and that way we'll be nearer for the return trip tomorrow. We can phone your father and he can organise it, so we hang on to this plane."

She left him, talked to the pilot and returned. "He thinks he can get in at Ambon. There are three or four good hotels there and he'll try to fix us up with some rooms,"

"Rooms?" queried Walton, looking at her. She smiled back. "They know who I am. We will have to be discreet."

Walton apologised. "I'm sorry for all this fuss, but it could be the breakthrough. I guess I sort of got carried away didn't I?"

She squeezed his hand. "All I want you to remember is that I'm part of it." He noticed the marigolds. She explained they didn't normally grow on the islands, but that particular area had a lot of connections with the Dutch and they'd planted some.

"There aren't many left now because when the Dutch left nobody bothered about them. Orchids came in and marigolds went out of fashion. But I like them."

Something stirred at the back of Walton's mind, some long-forgotten fact he'd taken in and mentally parked. It had something to do with the Dutch and pines. She agreed he could have a couple for his bag of samples.

They landed and took a taxi for the thirty-kilometre trip from the airport to the Hotel Mutiara, leaving the pilot to make the arrangements for the morning.

Over dinner Walton quizzed Lestari about who might know more about the island. Someone must have planted the marigolds, for instance. Did the military, he wondered, keep any check on who went to and from the islands? Were the tree-men living there now? He could hardly wait for the morning.

Lestari was quiet, also puzzling over the problem, but still she had the feeling it was only a matter of time before Walton went back to America. Would he ever return? She wondered about volunteering to go with him, but who would look after her father?

Eventually he caught on to her mood and teased her about whose room they should sleep in – they had managed to get adjoining ones.

She smiled but glanced nervously round the dining room. "You'll ruin my reputation, it will create a scandal and my father will not thank you for it."

For a moment he was aghast. "Sleep alone? I couldn't."

She insisted. He needed lots of rest. And after all, she said, he had to sleep alone when he was in America didn't he? Walton didn't even blink a response. He agreed that was so, but this was different. He couldn't adequately explain it but assured her that it was.

It was as well that he did sleep alone. He spent most of the night, dozing fitfully, excited by the thought or a possible solution, getting up in the middle of the night to glare again and again at the map. There were only two words that gave a clue and they were in Indonesian. He wondered what they meant.

Over breakfast Lestari translated. "That's nice" she said *"The island with flowers.* That's about the only translation -- though there's an inference that the flowers are strange or golden."

"Like marigolds?" he asked. She agreed it could be. Walton still had a gut feeling about the flowers.

When they reached the island Walton dashed ashore, heedless of any danger. He found the marigolds, a carpet of them at the head of a small stream, He paused. He could hear birds, the normal sounds or jungle life, the place was crawling with insects and here and there the carcass of a PWN. But what had wiped them out? Everything was as you would find in a normal jungle. Everything with the exception of this place; this place had marigolds. It had to be a clue; at the very least.

Within an hour he was hustling them off the island again and back into the plane. He was increasingly certain the marigolds had an answer for him. As they took off Lestari noticed two small boats on the other side of the island. It had to be the tree-men. They'd found their refuge. Walton was glad for them, but his more urgent concern was the need for a library and a phone.

They returned to Ambon, just made the connection with the internal Garuda flight and by that evening Walton was telling Lestari's father the news.

The old man knew little about the area near the island. He could only recall the Dutch had been strong there in the past. And, he remembered now, it had been a great area for flowers, even roses in the past he'd been told.

Walton sat bolt upright. That did ring a bell, but the old man, despite Walton's questioning could remember no more about it.

Nani came in then, complaining they'd been away two days instead of one. "It was a long time" she pouted. Walton promised her a bed time story. Together he and Lestari put her to bed, Walton telling her a story, making it up as he went along. It meant they missed the nightly Voice of America news, but the old man told them about it when they returned to the room. Lean had offered a ten-million-dollar reward for anyone with the solution to the PWN problem. Walton perked up "I could get by on that."

Soon they were debating what changes a solution would bring,

The old man had a theory that America might be better off eventually, rediscovering much that had been lost. "But what about Indonesia?" asked Walton. The old man surprised him. He wasn't sure. In the past they'd exploited the trees as a cash crop, now perhaps they could learn to use the land for food for the people.

Walton agreed he had a point, but the trees were still desperately needed to provide a "sink" for the absorption of carbon dioxide. "But don't other nations have some responsibility there too?" asked the old man gently. "Why should it be us that covers our land with trees so that others, wealthier than us can devote their land to the growing of food?" Walton saw what he meant. It was a global responsibility.

The old man looked at Walton and -- to Walton's surprise -- at Lestari, "I think you need your sleep; you have much work to do in the days to come."

He bade them goodnight and Walton left for the cottage. An hour later Lestari slipped joyfully into his bed, thrilled by her father's acceptance that she had a new man in her life.

Walton was fast asleep and snoring. She kissed him gently. On his pillow she left a marigold. At the moment, she felt, that was closer to his heart.

## CHAPTER SIXTEEN

It had only been through Grace's persistent calling they'd managed to get a room at all. Kansas was bursting at the seams.

Powell knew the locals were touchy about it but the old remark about the town being "a one-night stand between Chicago and the Rockies" was becoming true once again.

Hundreds of couples, it seemed, had taken to the idea of Operation Pioneer Spirit, left the kids with Mom and decided to look at the ground for themselves. They'd flown, driven and even poured in by Amtrak. The place was bustling, reminiscent said the locals, of the gold rush days. And it somehow had that air about it. Powell knew now, he'd been right to talk his editor into another venture out on the road. He hadn't stressed he was taking Grace along for the ride, but there was little difference in the price, and it was about time they'd got out of their routine.

Eating that night in Houlihan's Old Place had somehow reinforced a link with the pioneer days. It felt like a Wild West bordello, but so what -- at least it wasn't Washington. And it was good to have Grace along. He'd never forgotten how lonely his last night in a hotel here had been. Now it didn't matter that he had nothing to read; there were better ways of spending the night.

He looked at her over his brandy. Tonight, there was a sparkle about her that augured well for later on. She was so much more relaxed here.

Gone was the gloom that hung over from long days dealing with the problems of the world in the White House. Tonight, she was a gal out with her man and determined to enjoy it. He dragged his eyes off her, suddenly remembering they had company. Ted Johnson was an old pal from way back when they worked together on newspapers in San Francisco. Like so many, he'd switched over to radio, moved East, married a local girl and settled down. The girls raised their eyebrows slightly at some of the tales but mostly shrugged them off as bravado.

Powell looked at his watch and then at Grace. She nodded.

Again, Powell checked the arrangements for the morning with Ted -- he'd begged a ride in the station's helicopter in exchange for doing his almost now routine of "A View from Washington" piece. It was a fair swap. Grace snuggled against him joyfully in the back of the taxi, slipping her hand down to his crotch. "I never knew you had such a wicked past," she teased "I was quite shocked."

He gave an evil sounding laugh. "Yeah -- and I have a couple of wicked hours coming up too."

***

Next morning, they barely made it to the airfield in time. "What kept you?" asked Ted. Powell grinned wearily at him. "You're too young to know."

They headed out over the Missouri, then north towards the plains. Powell noticed how brown the river was and wondered just how successful the dredging operation was. Literally as far as he could see the plains rolled out towards the sky but broken now and then by small shacks. He asked Ted what they were.

"The newcomers" he shouted. "They're all over."

After half an hour they dropped back into the city, noting how packed the freeways were. It was a long ribbon of campers and trucks -- all heading west, and not all of them turning off for the city. It really was the great trek west all over again.

They dropped Ted at his office downtown, got into their hired car and headed out again towards the plains. Powell knew he had to talk to the newcomers -- and the locals for that matter. Twenty miles out he took a right and kept going until the road deteriorated into a dirt track. At the end he could see one of the new breeze block shacks, a camper by its side. Over in one corner of a small field Powell could see a man, hunched over a spade. The woman came to the door, arms covered in flour. He introduced himself and Grace, accepted the invitation, and they went inside for coffee.

The woman said it wasn't like the place they had back East, but they could eat here. They were living on their savings right now, but come the spring, if it all worked out, they'd have their own vegetables and maybe after that they'd have some wheat to sell too. The woman was eager to talk -- it was one thing she missed. She'd often go days without speaking to anyone and the landlord farmer treated them like they were shit. They paid him good money, and in advance, but part of the deal was that they had advice with it. That had been in the agreement. When her husband asked about that he'd got it straight. "My advice is you learn the way I did -- you make your own mistakes." It had been that simple and they'd barely seen him since.

But that had been early on. Now, her husband, when he didn't fall asleep exhausted, would drive into the nearest town and join the others for night classes. "Basic husbandry" they call it, and it had become a joke. At first the men had been so tired they didn't have the energy to be husbands any more. But now it was better. He'd lost 20 pounds and was more fit for it. She couldn't remember the last time he'd had a flat stomach.

The kids too in many ways, were now getting used to it. They'd complained a lot at first and the one who was away at university couldn't 'phone them anymore. It was hard, but somehow they coped. The youngest who, in the old days, would have run from a city poodle was now the family expert at milking. Not just the cow, but the goat too.

"And you know what?" she asked Powell "Me and my old man we even get to talking in the evenings. He sleeps when he comes in, but not so much as he did at first. And out here, somehow, it's a better kind of sleep."

After coffee, they strolled across the field to the husband, leaning on his spade and mopping his brow. The new metal shaft glinted in the heat of the sun. The man spotted Bob's recorder and his wife explained who he was. The man had little to say. Bob persisted, asking him how much of a change it was from his old job in a factory.

"You paying me?" asked the man. Powell hadn't thought of it. It wasn't normal. "You heard of barter -- it's real big out here now."

Powell had heard about it. Ted had told him last night about the new barter clubs.

The man gave out with a slow broad smile. "You want information from me, there's a fee. If you don't pay cash, you dig that row for me, and then I'll talk to you."

Powell heard Grace snigger beside him. They were all watching him. There was no way out of it. He laid down his coat, put the recorder on top and grasped the hot metal handle of the shovel. It took exactly 28 minutes. Grace timed it.

While he dug they went back to the house for a cool lime juice.

***

Bob knocked, walked in and grinned. "You know you won on that deal. I don't need to ask you a thing. I just know what it's like, every muscle in my back is telling me."

They laughed. But Bob did have one question for the man. "At the end of that field the ground rises right?" The man agreed.

"But there aren't any marks where the water usually brings the soil down with it. Now how is that? It almost looked as if there was a terrace underground there somewhere."

The man said there was -- a prime example of old-fashioned American ingenuity. He looked over at Bob "You know the guy who thought of that came out here just like me. He didn't have a lot of money and he got to thinking about all that erosion stuff. He'd seen the President and thought that way was kinda crazy.

Now the guy was a drinker see and when he came out he brought a dozen bottles of Jack Daniels with him. Real fond of that he was. He brought it out in one of those old plastic milk bottle carriers and threw it away in the yard. One night it rained and next day he saw it had stopped a lot of silt in the yard. So, he played around with it, put some wire mesh on the bottom, dug a little trench and put it in the ground where the silt normally ran down into the yard. Stopped it dead. Didn't have no more trouble at all. Well it was a long story, but he got some money together and now he makes them all the time. Even has guys working for him now. They only cost two bucks and there's not a country in the world where they can't be sold. Any old peasant in any old country can work it. They reckon this time next year he'll have made a million bucks."

Powell listened with delight. Exactly the sort of thing he wanted. He stood up. "It was almost worth digging that row."

He went back to the station, recorded a piece for his editor and talked to him about the story, asking if he could put in for a new back on expenses. The editor pulled his leg about maybe exercising it less at nights and agreed he could stay another night.

***

They went for lunch downtown. As usual, he couldn't help hearing the conversation at the next table; a chat between two real estate agents glorying in their new-found business. All the sort of rackets he'd suspected were going on actually were happening. They were in on all the concessions for butane bottles, water heaters, even kerosene supplies. Grace looked despairingly across the table at Bob, chiding him silently. He nodded distractedly, concentrating hard on the conversation.

The men left and Powell looked at Grace. "Did you hear that?" She confessed she hadn't -- she'd been deafened by the sound of his brain working overtime. He told her what he'd heard and wasn't it sick how some people would profiteer off of anything.

She was not so concerned. "You can't stop it. It happens. And it'll soon settle down." Bob supposed it would, but still thought it vicious.

He called Ted at the radio station to see what entertainment there was in town. He was astonished. There were at least five live concerts -- big name stars now back out on the road since records had more or less vanished; and the tickets were expensive, but Ted had a couple of handout ones. Bob was welcome to them.

***

They spent the afternoon playing tourists, wandering through the Crown Center, Bob explaining how before the crisis there was a tropical forest inside. Even in the last two months the whole aura of the place had changed dramatically. Gone were the chic fashion stores. Deserted furniture mausoleums were now were shops stuffed full of agricultural tools, seed shops and good old-fashioned farming clothes; mostly cotton dungarees.

Real estate agents had arrived in a big way and a bookstore was now converted into an advice centre run by the Soil Conservation Service. It was packed.

Bob had always been told there were two kinds of shopping -- fantasy and workaday, but now it was all one.

Every shop was strictly practical, and each had a feel of urgency about it. Where people had in the past stopped at shop windows and dreamed now they bustled in and demanded answers. And, often, if they didn't have money they'd haggle a barter. Powell grinned at Grace

"You realise you could swap all your blouses for a set of gardening tools?" She teased him back "It might almost be worth it to see you digging another row."

He groaned, the stiffness was still with him and when he'd told the editor he'd just laughed as well. They flew back the next day, making their descent to touchdown over the now bare, dull-green swamp where a mini forest had once marked the final approach to Washington's airport. They picked up the car and drove back to the flat. The roads were almost empty and downtown had the feel of a ghost town about it. Here and there a boarded-up restaurant bore testimony to the ever-decreasing number of people who were in the city. Powell looked at it with sadness. "Just think" he told Grace "this city is probably better off than almost anywhere, bar New York. I mean stuffed full of diplomats and all the people who go with them. And look at it!" She agreed but reminded him the American Government, itself, now had far fewer people in the capital. So many had been seconded out to State capitals or their jobs had simply been made redundant.

That was what had made the difference. But the ghettos were emptying fast as well, the jobless blacks joining the great trek west. How long, Bob wondered, could the city even support itself. Maybe it was something he could do a piece on. It was about time he talked to Joe Simon again -- maybe he'd have some thoughts on it.

***

He mentioned the idea at the conference next morning. At least it was a way of getting them off the theories about how he'd ended up with a sore back.

"You mean are we seeing the end of city life -- something like that?" asked the editor. Powell nodded. The editor said fine do the piece -- and upstairs would be interested too. With fewer listeners the sales department had to drop its rates and that wasn't good news. For anybody. The editor glowered -- looking at Casey. Casey, still terrified of the man, gulped. "Yes sir?"

The editor growled. "Noon plane. Detroit. General Motors first mass production electric car. Right?"

"Yes sir" said Casey, acknowledging it was the only brief he'd ever get. He left and halted in his tracks as the editor growled again. "And Casey."

"Yes sir".

"Casey, this time why not take your recorder with you?"

"Huh?" Casey blushed and stumbled away. The reporters chuckled. They'd all done it once in their lives and yesterday it had been Casey's turn, running out the door to the scene of a fire, then ringing in half an hour later to ask if someone could get his recorder out to him in a taxi.

The conference broke up and Powell went to his phone and called Joe Simon's office. Lately it had been a lot easier to get him direct. A private number only interrupted by his secretary. He'd jotted it down when he was left alone in Simon's office once.

Simon had heard his piece from Kansas and he too was joking about his back. He knew Grace had gone with him. He knew because Walton had called, and her assistant had asked his secretary where she was. It was the way the White House grapevine worked. The thought had made Simon uneasy for a while, a reporter having an affair with the secretary to the Scientific Committee. But, he reasoned, it might force Powell into discretion rather than indiscretion. Things often worked out that way.

Simon too had been thinking about the downtown area and, by coincidence, had talked to Lean about it. Powell went straight round to catch him before lunch -- and got a surprise.

The Vice President explained that they were considering an idea for a state funded rehabilitation scheme already. The idea would work this way. Because people had moved out, all going at once the property market downtown had been glutted and prices had crashed. The government wanted to renew those areas, but like private enterprise, they didn't have the cash. Since the state didn't have to make an immediate profit they were going to issue a special bond to those who sold them their houses. Those couldn't be cashed in for ten years, but the owners wouldn't lose. That way the city centres could be renewed.

"But who'd rent or buy places downtown" asked Powell spotting what he thought was the weakness. "Rent or buy what?" asked Simon.

"I'm talking about parks, leisure areas, making city centres places for people to relax and enjoy themselves in. What law says you have to have office blocks downtown -- they've been moving out for years anyway."

"What I want to see. in years to come. is the business quarter grouped around a central park. You ever been to Oregon?" Powell had to confess he hadn't.

Simon explained "We've got a place there called Washington Park. It's 145 acres overall. It's got a zoo, it's got Rose Gardens and there's a museum too. Now why can't every city be like that?" He went on "We've got a great chance right now to do something good. We've got to use it. The way I see it everybody wins."

Powell switched the subject, knowing Casey was in Detroit, to what was happening there. Simon was enthusiastic to comment on that. "You know in my student days I led a march on the local General Motors office. There'd been a rumour that they'd got hold of a battery so that they could produce an electric car right then. Well when you're young and you're in Oregon you believe these things. As it happened it was just a rumour, but now it's actually happened and they're only the first. I don't think I'm giving away any secrets when I say they could have some competition soon. I think we're in for a very exciting time. You maybe know already that they've done a test run clear across the country, re-charging at points along the way. Well, we're the leaders in that technology and I can tell you there's a world market for it and with the dollar the way it is I think we're going to find a whole new export field." He paused, wondering if Powell wanted to ask a question. Powell quickly shook his head, extended both palms towards him in a signal to continue. Simon was underway now, enthusiasm taking over from political caution.

He went on "I heard you talking about the guy in Kansas with that idea about milk containers to stop silt. Great idea, and the guy deserves to make a million. That's how this country became great anyway. And I tell you Bob this country of ours is bursting with ideas. You know we started the Good Ideas bank -- I think you were the first to hear of it. Well every day I open that post personally, and that's just about the best start to a day a man could have. Just take one instance that came in today. You know we chucked out a lot of our old paperwork. Now an old man who used to work for the Government as a clerk knew darned well that all used to get taken away and stored. I didn't realise it either, but we had whole warehouses full of dead files. They all got cleared out and they were standing around empty. So, this guy writes in and says why don't we hand the empty warehouses over to local organisations and let them use them as sports halls. And why not -- I'm sure they're much more useful that way."

"And nationally, environmentally?" asked Powell, knowing Simon had a point to make. Simon gratefully, acknowledged the hint.

"It's getting better all the time. Pollution is way down, LA hasn't had a decent smog for months and the incidence of heart disease is the lowest for ten years. Next month we crop the first papyrus and I'll let you in on a secret here. It isn't definite yet, but we may well use the first run of that new paper to print ourselves some brand new dollar bills."

Powell sat back, ending the interview. He had more than enough. He grinned over at Simon. "I'm beginning to believe you like the country better this way."

Simon paused, taking it seriously. "You know Bob, you could be right. A lot of people have suffered, and the climate thing still bothers us; and heaven knows how they're coping out in Indonesia, but in purely domestic terms it could be the start of a whole new American dream."

"You mind if I use that?" asked Powell, "The new American dream bit?"

Joe grinned, "No just pay me the usual fee."

***

Powell thanked Joe and left, dropping in to Grace's office along the way. She wasn't there.

On her desk was the now routine deletable chinagraph message pad. She'd not cleared it and he could see -- without even twisting his neck too much -- that Walton had called.

"Told him you were out of town" said the message. He noticed that the words had now been crossed through. He wondered if she'd returned the call. As he finished scanning the pad she walked in.

"Yes. I did ring him back" answering his unspoken query. "He asked me how life was in Kansas."

"How the hell did he know that?" asked Powell.

She explained: because she wasn't there, Walton had called Joe Simon's secretary who'd let it slip. Then he'd heard Powell's piece from Kansas repeated on Voice of America and put two and two together.

"Does it bother you?" he asked quietly. She shook her head slowly, walked over to him and kissed him gently.

"Not now. A time ago maybe, but not now."

It was a subject they'd hardly talked about before. It still had the feel of taboo about it.

"Quite certain?" asked Powell.

She kissed him again. “Quite certain."

## CHAPTER SEVENTEEN

The laboratory was quiet now and Walton was left alone, hunched as he had been for hours over a microscope. Around his slide was a litter of stems and petals; on the bench behind him, test tubes and more slides, some containing wood boring beetles, others filled with the sticky gum of tylosis. He was in effect doing a post mortem backwards, guessing at a cause of death and now trying to repeat the process.

Here and there were small containers holding live PWN.

In his own mind he was sure there was a connection, sure that somehow the presence of the marigolds was in some way a cause of the PWN's death. But what was that link, if there was one?

He went through it again. The acid took the leaves from the trees, opened up wounds allowing the beetles into the trees. There they bred, emerged and were seized on by the tiny PWN worms. At no time did they go anywhere near the marigolds. Yet their presence had been the only thing that differentiated that island from the rest. It didn't make sense -- unless there was something in the marigolds that killed the worms. But how did they come into contact with each other?

A connecting link - that had to be the logical answer.

So, he argued, he had to find that link. Or did he? Was that really his prime objective? Of course, it wasn't.

What he wanted to do first was to find a way of killing PWN and he thought he had a clue to that -- the marigolds. But that didn't mean he had to do it the same way that nature did. After all, pesticides didn't crop up naturally in nature. Man found out what killed something, manufactured it and then sprayed it on directly. On that basis he didn't need to find the link.

The prime question to answer was "did something in the marigolds kill PWN?" He took one of the live PWN and placed it carefully in the slide. Taking a test tube, he gently placed a drop of the liquid from a crushed marigold into the same slide and went back to the microscope.

Would it die? Or would it live? That was the ten-million-dollar question. He'd been thinking about that ten million dollars reward offered by the President all day. Oh boy what a time he could have with that. Gently he prodded the PWN to get it moving. It stayed inert. It looked dead. Walton gulped and prodded it again. Was there any sign of life or was it merely comatose? How the hell did you tell if a worm was dead? He flicked it over, it was almost stick like now, almost as if it were paralysed. He left it a moment, walking back to the other side of the room, to check on his remaining stock of marigolds. Lestari had delighted in those and he still felt guilty about hijacking her whole bunch. He flicked his hand in annoyance.

A small ant fell off and scurried back into the bunch. Walton was irritated. What sort of laboratory was this? Fancy letting ants into the place. He walked back to the slide. By every sign he could imagine the PWN had died. It was stiff and utterly lifeless. He had to re-check it. That one could have died from something else. It could have been a freak.

For hours he worked on, the sky now lightening as dawn approached. Time and again he went back to the bunch of marigolds, carefully crushing stalks and petals in turn into a liquid, then using the liquid -- and he was sure it was that now -- to kill the PWN.

There weren't many of the flowers left now, he made a mental note to have some more collected. The army would probably think it odd being asked to go out and pick flowers, but maybe Lestari would do it for him. The army was used to acting as unpaid servants to Government Ministers out here -- they'd just put it down as another example of them living the high life at the country's expense.

He gathered the last flowers in his hand and studied them as he walked back to his slides. He laid them on the bench behind him and went back to his microscope. Again, with annoyance, he flicked an ant off his hand. He'd have a word with those bloody assistants when they turned up. They were so lax it was unbelievable.

For another quarter of an hour he bent over the microscope, carefully peeling away a thin layer of the PWN.

It told him nothing, he'd try it on a live one and see if he could notice any difference. He turned to the bench to pick up another PWN. He was aching now, the muscles in his neck getting stiff. And he was getting edgy. Where were the live PWN? These were no good, they were all dead. Where had he put the live ones? He knew he'd put them near the flowers. And there was another ant, and another, clustered round the dead PWN. He quietly sat down on his stool, stared at them and laughed softly to himself. He was looking at the link. The ants were transmitting a substance from the marigolds to the PWN and killing them. Of course. Ants always had used poison of some sort to stun their victims, then whole armies of them would carry away the body for food. Why the hell hadn't he thought of that before? It was so bloody simple. He took the last two flowers and crushed them down into the liquid he wanted and transferred it to a test tube. He sealed the top, packed it into a metal carrying tube and slipped it into his pocket.

For the next half hour, he was busy destroying all traces of what he'd been working on. He'd deliberately worked totally alone in the laboratory for security's sake. There was no way anyone was going to steal his idea.

He swept all the flowers, the stalks and the PWN, and what ants he could find into a container and carefully burned the lot, sweeping the ashes down the sink.

He meticulously washed down the slides, put all the equipment away and after a final check left the laboratory. There were no notes, no clues, just a tube in his pocket. He drove blearily through the dawn, occasionally patting his pocket as if to convince himself it was still there. He went back to the cottage, left a "don't disturb" notice on his door and slept ‘til noon.

***

He walked in the garden before lunch, racking his brains about the next moves. He wondered what the substance was; he'd not had a chance to analyse it yet. He worried over whether it could be synthesised. It was all very well the ants transferring it in the wild of the jungle, but maybe marigolds wouldn't grow everywhere that trees did. Still he wondered where he'd first heard about marigolds. And there was that mention of roses. He still had a lot of work to do. He mentally reminded himself it had to be in secret.

The sample he had in his pocket was barely enough for a proper series of experiments so the first step would be to get hold of more samples. Lestari could fix that for him. As a double check, he'd get hold of some of the ants as well. He was sure in his own mind they were merely carriers because the liquid from the flowers had killed the PWN anyway.

But maybe that liquid only worked when it had some contact with ants. It was just conceivable that they transferred something else also to the plant, and it was a combined mixture that killed the PWN.

There really was a lot of work ahead. But should he do it here? Again, the security problem was double edged. It was quieter here, but the more he was seen to be using marigolds and ants and PWN the quicker people would put two and two together. He couldn't really involve anyone else. He thought through it all again, then decided.

He'd do all his basic work here in assembling the materials he needed for analysis and, subject to what it was, see how the substance could be produced in quantity. Then, using another laboratory he could use the manufactured substance for final field tests.

Lestari interrupted him, reminding him it was time for lunch. She noticed his distracted look, saw the bags under his eyes and asked him if he was alright. She'd noticed the sign on his door and presumed he'd worked through the night. He held her hand and quietly told her he thought he might be getting somewhere with the marigold theory.

She smiled. "I hope you don't mind, but I've been doing some research this morning as well."

She'd discovered the link between marigolds and roses and nematodes. She explained "The Dutch in Holland years ago had problems with another sort of nematode that was destroying roses. They, too, had discovered that if they put the occasional marigold between the bushes the nematodes died or went away."

Walton nodded. It was near enough. It wasn't quite the same because PWN in this case had the killer substance transferred to them by ants. But he had heard of it before and it was good to know the common factor was the marigold. She had more information.

He teased her "You have been busy."

She smiled. "I found a better translation of the two words on the map for the island. They meant the island of the golden flowers --it was in a slightly different dialect and the words on the map were a corruption of the original. The golden flowers must have been the marigolds."

Walton nodded thoughtfully -- he supposed it was inevitable that Lestari and her father at least knew he was near to a solution. He didn't suppose it mattered too much. In a way it would help, because he would need facilities and assistance. He told her of his need for more flowers and she said she would arrange it. She wouldn't go herself, that would attract too much comment and attention. It would be in the form of a straight requisition from the Forest Service, but the samples would come straight to Walton.

***

They went into lunch. Lestari's father looked at him quizzically. "You work very hard for us. We appreciate it." Walton filled him in in very simple terms on what he thought he had found. He told him of his plans to do stage two of the research in Portland and stage three at a third laboratory. For some reason the old man was insistent that stage three of the research should be done in Indonesia.

Walton queried it and the old man explained "Quite simply I think that apart from yourself of course the solution is more ours than anyone else's. I suspect our need is greater than America's."

Walton couldn't disagree. He looked up at the old man. "When we last talked you had some doubts about making the discovery public. You thought, as I do perhaps, that maybe both your country and mine are better off in their new situations."

The old man agreed "I thought about it a lot and I am still considering all the aspects. But, as I mentioned before, it does seem that the land now available to us should perhaps be used in large areas for growing food for our people. For many years now we have simply been exploited as a nation by international corporations for the timber we had. I do not want that situation to be repeated. Of course, I want us to have timber and we have a responsibility in that area; I acknowledge that. But is it the duty of a poor nation to lay itself open to exploitation for the benefit of richer countries? I think that is the crucial question. Your solution, if indeed it is one, would, when fully applied merely put us back to what you call square one. I realise I am talking about a very long-term situation, one that would continue long after my death. For too long we have ignored our responsibilities towards those of say, Nani's generation."

"And what about America?" asked Walton.

The old man paused. "In a way that is not my concern. All I would ask is that you think very carefully and consider the things that have happened there in recent times. From all the reports we have, the national attitude has changed in many ways; there is less waste and less obsession with material matters. It is of course, not my concern; it is yours, and in a way it is yours alone. You have the temptation of a huge reward, of national fame, a life of ease and a chance for a place in scientific history; all of which must have their different appeals for you. But you also have a grave responsibility. Think ahead. Your solution could bring back the America of old. I am confident you will think deeply on these matters."

Walton said nothing. Nor did Lestari. The old man left them. Lestari slipped her hand across to his. "Dick. You know I love you and so does Nani. Whatever you decide I'm sure will be right. But please don't forget that we love you."

A small tear stung Walton's eye. Nani -- her future and of generations after her. But she had no say in the decision to be made. He had to make it for her. And Lestari too, dear sweet gentle Lestari. He gulped, and silently patted her hand, gruffly saying he wouldn't forget.

She left him then and he went to walk in the garden. At least he would have some time to think.

The samples would be back tonight, by morning he'd be on the plane for Portland and in a matter of days he'd be back. Planes were always good places to think and he needed all the help he could get.

Again, that night he worked late, quickly distilling the flowers into a liquid, and as a precaution, some of the ants as well. Once again he carefully destroyed all traces of his work and wearily, this time with two tubes in his pocket, made his way back to the cottage.

***

Lestari was waiting for him. In bed. He undressed and gladly slipped in beside her. She held him close and whispered to him. "There are times when it is better not to think." With an infinite, yet sensuous gentleness she started to make love to him, smoothing his body with her hands, then her lips.

Walton's tiredness slipped from him and now it was his turn to explore her. Soon the pace quickened, and he was taking her now, gratefully letting his body take over from his brain. She understood it and now her body too took over, a gasping urgency that soon left them shuddering and spent.

He slept, exhausted in mind and body, and woke relaxed and fresh. No matter how many times he did it, the excitement of a trip never failed to fill him with a flowing energy. Lestari noticed it and on the trip to the airport, merely slipped her hand across to his and squeezed it.

***

The farewell was formal, as it had to be, and soon he was sitting slumped in his seat above the clouds. He looked down on them and, not for the first time, the thought came to him that this was God's eye view of the world. Yet wasn't this what he was doing. Playing God? Was it his function to make the big decision about whether America should be granted his solution?

Wasn't his function purely to work as directed and produce the result for Joe Simon and the rest? At least there was no decision to be made that way -- but that in itself was a decision. What, he wondered, was his responsibility?

His mind drifted back to university days and all the into-the-night student debates they'd had then; most of them stemming from defense projects. Did the scientist have a responsibility at all, and if so, what and to whom? It was a never-ending discussion. But now he had to decide.

The air hostess smiled down at him, replacing his drink. There was something of Grace's chic about her. And that was another question altogether.

These months with Lestari had been idyllic and instantly his mind went back to visions of her naked on the tropic beach. But would that sustain him all his life? Didn't he need the pertness of Grace's mind, the discipline of having to treat her like a lady. At least she never let him become complacent.

The more he thought of it, the more he realized she had to be the better long-term solution. And now he had the chance to capture that prize. He could see himself now, the king of the nation and her as the bonus. But wasn't she more than that too? There had been so many good times, so many occasions when he'd just dropped his jaw and looked at her in wonder. Or listened to her. She was a beauty and there was no doubt she had one of the best minds he'd ever encountered. She was sharp; she had judgement and integrity. She always had good points to find in people. And she was of his culture, of his background. Surely that held the promise of greater stability in the end.

He idly wondered what she was doing now. He'd call her when he got to Oregon. Even though the last time he'd spoken to her she'd been cold, quite chilling in fact. There was no doubt in his mind she'd been away with Bob Powell. But who was he compared to a guy who would give America back its life?

No, the weapon he had now was that mysterious aura of power that so attracted women in a way he'd never understood. *The man who gave America back its life....* a nice epithet.

***

The plane landed and he was pitched into the bustle of the airport. No doubt about it, America had a unique energy unknown anywhere else on earth. It was good to be back home. And in Portland too where there was so much sympathy for the ecologists, and, yes, he still thought of himself as one.

An ecologist... somehow it had so much more of a caring ring to it than scientist.

Joe Simon thought of himself that way too. Maybe, Dick thought, he could talk to Joe. Come to think of it Joe was about the only man he could talk to.

In a way the whole damned thing was a nuisance. He hated coming to a decision of any sort and he knew he was going to have to finally decide between Grace and Lestari. OK so Grace wasn't his, at least not at the moment, but she could be. He knew if he really worked at it she could be. But the other thing, well that was a whole lot bigger.

He drove out to the laboratory, deciding he didn't have to decide about that right now. First he had to be sure of the PNW solution; that his theories could be turned into fact. He felt safer in that territory; it was straight-forward, and the facts made up his mind for him. He checked his watch. 5 p.m. here meant it was 8 p.m. in Washington. She should be home by now. He'd call her. At the very least she should know which country he was in.

***

He arrived at the laboratory and rang her, at the flat. A man answered the phone and asked who it was. He paused, thought about it and quickly put the phone down. 8 p.m. and he was in her flat already. He'd been sure it was Powell; that carefully modulated broadcaster's voice was unmistakable.

The very fact that he'd answered the phone proved she didn't mind people knowing he was there. He knew how naturally discreet, almost secretive, she was. It could only mean he was living there, and it was as near as, dammit, public knowledge.

And how about her discretion over work?

There she was, secretary to the Scientific Committee -- which nowadays had to be handling classified material -- going around with the biggest mouth in Washington. A sudden burst of rage spread across him. How dare she? All that crap about how she valued loyalty and the rest. Why he'd only been away... well it was a month or so now, but even so, hadn't he called her most days? OK so they'd had a slight fight as well, but for Christ's sake did she have to take it to that extent! And that bastard Powell, he could only be using her. He wondered if she had any idea about that. He'd have to call her tomorrow and warn her. He pulled himself up short. If he hadn't known himself better he'd have diagnosed jealousy. The thought of her with Powell sickened him.

He thought for a second of driving back to the airport, getting the evening flight and going to Washington. He'd arrive in the middle of the night, but what the hell. If Powell was there at least he'd know about it, there wouldn't be this agony of not knowing. That was the worst bit. Maybe he should call again in a couple of hours, or three.

***

The night duty laboratory assistant put his head around the door to see if he had everything he needed. Walton grumped he had. The lab assistant told him where the coffee was and left him to it. Walton looked around at the facilities. The station had been set up to be part of the increasing effort put in by the forest service into research and development. He knew they'd had problems with budgets over the years, but now they were in the front line and, for the first time in their history, facilities were available on demand.

He set to work, his mind was clearer in the night and soon all thoughts of Grace disappeared as he became immersed.

He soon established it was an alkaline substance and then, painstakingly, went through a series of tests to determine its likely groupings and place in the elemental ladder. It was easier than he thought, why they even had some textbooks and manuals left. What an exhilarating feeling it was to be able to read again.

He'd not realised just how much he'd missed the printed word; how it stimulated and expanded the brain.

If he was right about all this, then in a few years people would have books again -- and gift-wrapping they could carelessly discard. Every advantage he thought of was countered now by some matching waste. Steadily he worked, through the night and the early hours, eliminating side issues along the way. It took him a long time because so many plants produced so many versions of alkaloids, and those were only the ones that had been discovered and properly tested.

He recalled a previous study he'd done for some medical research and he'd been fascinated by the number of commercially produced drugs derived from plants. There were so many examples, strychnine, cocaine, morphine and nicotine and the hallucinogens like LSD and mescaline. As he recalled, there were something like 300,000 flowering species and so far only around 2 per cent of them had been tested. Yet even from that 2 per cent had come drugs to treat cancer and heart disease, leukaemia and schizophrenia. Almost all of them, too, he recalled came from the tropical moist forests; the ones ecologists said were being destroyed for cheaper hamburgers.

Time and again fellow scientists and groups like Earthwatch had screamed that the other 98% must be tested; that they could contain drugs of immeasurable benefit to man. Yet nothing had happened.

Just something like twenty bulldozers, given a week could have ripped through the part of Mexico that bred the vine that gave most of the world its contraceptive pill. It was only by sheer luck that it hadn't happened. Yet it could have happened and the world, unknowingly, would have destroyed arguably one of the most important drugs ever known.

The whole thing was crazy. He was getting tired and fanciful, bitter and vengeful against so called civilisation. Did it really deserve his effort at all?

An hour later he was sure of what he had. A substance that relatively cheaply and easily could be manufactured and sprayed on forests to defend them against PWN. Another insecticide against another disease. On the current count that was one more to add to the 58,831 already registered under commercial names. Such was his contribution to mankind. How ironic that for this one they were willing to award a prize of ten million dollars.

He was exhausted now and slowly he cleared away all evidence of what he'd been working on. He could memorise the formula and needed nothing else. He washed all the equipment clean, sloshed his compounds away down the sink and left. He'd return tomorrow and make up a sample spray he could test in Indonesia. That way he'd be the only person in the world to know exactly what the answer was.

He drove back to the city and slumped into his hotel room, the formula buzzing around in his head. He was almost dizzy from exhaustion. He looked at his watch, setting the alarm for 10 a.m.

***

It was 4 a.m. local time and in his mind he had automatically converted it to 7 a.m. Washington time. He picked up the phone and dialled Grace's number. A man answered.

Walton disguised his voice. "Jim?" he asked.

"No, it's Bob" said the sleepy voice.

Walton replaced the phone. The bastard. There last night and still there at 7 a.m. How could she? The bitch! How could she? He could have wept. Instead he just slept.

By noon he was back in the laboratory and by 5 p.m. was on the flight back to Jakarta, a bulky cylinder of insecticide with his baggage. The customs men had been awkward for a while, but eventually the UN pass and the credentials had satisfied them. They'd asked him the value and he'd said ten million dollars or nothing. Sure thing, they said, thinking it was a joke, and passed him through. It was almost going home time for them and they had a feeling Walton would cause trouble if he had to. And who needed it? Not them.

Customs at the other end were worse -- until he gave them Lestari's number to call. After that it was all deference and he was waved through.

He'd slept in snatches on the plane, but still he felt drained, almost light headed with the difference in the time zones. His stomach was in turmoil and it was all he could do to slump, weary and aching into the cool bed of the guest cottage.

***

He woke around ten, then dozed luxuriously through until two. Lestari had left a kettle and coffee and milk by his bedside in the night. Deep in sleep He hadn't heard her come in.

He took his time over his coffee, opened the shutters and winced at the blazing sun. In a far corner of the garden Nani was playing with the doll he'd given her. Three years old, full of wonder, a magic of her own, totally lost in a world that was all hers.

Just a shout and he knew she'd come rushing over, cuddle him and want to know all over again that she was his very, very best girl. And in so many ways she was - in so many ways. Not having kids of his own he'd often envied the feelings on men's faces as they'd looked at their daughters.

It was a dreamlike gaze in many ways, an affection unsullied, unselfish too. And so protective, that mixed in with pride and just the feeling of wanting to hold the harshness of the world away from her. It was Billy Bigolo on the beach in Carousel, it was Sinatra's tribute to his Nancy, it was mawkish, sentimental and yet at the same time there was the element of pure, almost reverent love.

Just once he'd asked the editor what it was like having your own kid around. It was the only time he'd seen his face soften so fast.

"Dick -- it's everything. Maybe it's just me, but I think its most other guys too. For me she's the person I can't let down. I have to try to live up to the standards she sets for me. I just wouldn't ever, ever want her to look at me and say, *I'm ashamed of you*. I tell you, every now and then, I'll do something or say something, and she'll look round at me so pleased and say, *Dad I'm proud of you.* I know that sounds sentimental, but I tell you just those few words keep you going for days. She says it's the same for her too. It's like you've got this one person in the world you have to treat right. Not like the wife, some of the time, but every minute of the time. You just wait Dick. You wait 'til one day you have a daughter and she's little and she's ill. You pray so hard, you just force God down into that room to make her better."

Dick remembered how he'd looked… a granite tough old boot of a face that suddenly got a glow on it just thinking of her. He'd shrugged it off afterwards and said nobody could afford to have them, that this gave him more ulcers than work and his wife put together, but Dick never forgot the way he looked.

Nani... she was three now. And, just maybe, a large part of the happiness she'd ever know was in his hands. Some of her questions had reduced her mother to blushes and her grandfather to giggles. What kind of world was she going to grow up in? Caring or exploiting, sharing or grasping.

"She's beautiful isn't she?" He turned. Lestari had been standing there watching him. He pretended to have an itch and brushed away a tear that had rolled down his cheek. He nodded, his throat full. Lestari knew his mind was far away and briskly went about the cottage, making his coffee and running his bath. She shouted from the bathroom. "I'll keep her occupied for half an hour while you have a shave and a bath, but I warn you she is determined you won't escape without taking her for a walk and telling her a story. Is that OK?"

He yelled back that it was and moved quickly into the bathroom. He glimpsed himself in the mirror, checking whether in the muddle of time zones he needed a shave. He couldn't remember. In the mirror he saw Dick Walton, still red-eyed. "You bloody fool -- what did you ever do to deserve a daughter?" It was a good question. But he had no answer.

## CHAPTER EIGHTEEN

After two days working with his spray in the jungle, Walton knew his solution was effective. There had only needed to be the merest hint of the alkaloid derived from the marigolds for the PWN to simply fall over and die. It was quite deadly, yet it seemed to have no effect on anything else. So, where marigolds wouldn't grow, his spray would be the way to stamp out the disease.

He flew back to Jakarta and, late that evening, saw Lestari's father alone. He briefed him on the results he'd got, though the old man was barely interested in the technical details. Lestari's father eased back in his chair and studied Walton's face. It was elated, there was excitement there too, but also a lot of doubt.

"You are wondering what happens now?" asked the old man.

Walton looked over at him.

"I wondered if you had come to any conclusions."

The old man paused again. Ever since Walton had told him the news of the solution days ago he'd done little else but consider the consequences.

He told Walton "The simple thing to do, the obvious way, is for us to replant as many trees as we can knowing they will grow here more quickly than most other places on earth. Before long we would be a nation of great prosperity. As you well know the quickest trees to grow are softwoods and it is those that are in greatest demand. We would be farming trees and there would be no shortage of customers for them."

"It's the obvious course to take" said Walton "with the exception of a couple of snags."

The old man smiled at him.

"Yes, I know. In these past months you have taught me much. I know it would be short-sighted to put all our eggs in the basket of one type of tree. One type of disease arrives, and all the trees are killed. However, what worries me more is the long-term policy, the dilemma, if you like, that all the countries of the Third World find themselves in. We have a crop, but the farming of it takes an expertise we don't yet possess. Also, to make the maximum profit from that crop we should export processed timber and not raw logs. We have tried this as a policy in the past and, frankly, it has failed."

The old man paused for a moment then continued, "We should be left, as we are now, with having to rely on the large, rich and expert companies from America and Japan to farm that crop. We are, if you like, merely renting them the right to take our timber and make money from it. There is very little advantage in it for us and there are those who would exploit us. We should, as I said a few days ago, be merely going back to square one."

Walton interrupted. "There is another snag as well, the effect on the climate. If you were to plant nothing but softwoods like pine you would be putting more carbon dioxide back into the atmosphere than if you planted the traditional hardwoods."

"We would then be doing what almost every other country in the world is doing. Am I right?" asked the Vice President.

Walton had to acknowledge that he was.

The old man continued "So, the world would want us to plant hardwoods that take so much longer to grow in order not to upset the climate of other nations. But what do we do in the meantime? How are we to pay our way, how can we meet bills for the food our people need? Do we starve in order that the farmers on your plains can make a profit from selling us their grain? Is that a good deal for us?"

Walton realised that the old man had been through almost all the options. It was a delicate juggling act, balancing the immediate needs of his own people, against worldwide ecological demands.

The old man emphasised his point. "Why should the world demand such a sacrifice from us alone? Why should not America be told that it has to grow both soft and hard woods? Why should not the world demand of South America that it halts the rape of its forests in the Amazon, which it is doing purely for gain?"

Walton interrupted him there. "But South America is like yourself. It is relatively poor, it too has sold out to the large timber corporations."

"So" smiled the old man "is it not the answer for the world to look at these companies and say to them that they have responsibilities other than profit for their shareholders?" Walton laughed cynically "Try telling that to the guy who runs a multinational from Tokyo. I know where his sympathies lie. He'll just laugh at you."

"At first maybe" and now there was a graver, sterner note in the old man's voice. Walton asked what he meant.

"Because we have been poor in the past we have been exploited. We have had no weapons and yet in many ways we have been at war -- mostly with poverty. Now however we can make a new start. It is our decision and ours alone what we plant -- softwoods or hardwoods. We can even decide perhaps not to plant anything at all."

Walton raised his eyebrows. "But the rest of the world will argue that it needs your wood -- isn't that rather like blackmail?"

"Did the Arabs not do the same with their oil?" asked the old man.

Walton was getting lost. "So, what are you saying?"

The old man made it clear for him. "I am saying that our first priority is food and now large areas of our land will be turned over to the growing of crops. There will be whole areas of our country which will be used to produce food for our people rather than profits for the shareholders of giant companies who have exploited us in the past. In many ways we have no option. If we do not do this our people will starve. There are many crops we can grow and, as you know, there are some that put back into the air almost as much oxygen as the tropical forests which were there before. So, we therefore fulfil two of our main responsibilities. Food for our people and oxygen for the world."

"But won't you grow trees at all?" asked Walton, rather horrified.

"We shall see" said the old man. "If, for instance, a company came to us saying that it needed timber we should agree with them that an arrangement could be made. You see we have time now. In the time it takes for the trees to grow we shall train our people to be ready for the harvest. There are UN funds available for such training. In essence, instead of renting out the right for people to take our crop, we shall farm that crop ourselves and sell it to whoever will pay the best price for it. It may be that we shall not have material luxuries for some time, perhaps ten years, but we shall put that time to good use. It is only a minute of time, only a moment's pause in our history. Since it has been forced on us we shall make the best use of it."

Walton had to agree it made sense. They needed food; they needed a new deal over timber because they had been ripped off in the past.

But where did it leave him? The question must have showed on his face.

"It means" the old man smiled "there is no overwhelming urgency for us to have your solution. It means the only decision you now have to make concerns your country. I think you have an expression about 'playing it cool'. It seems sensible for us to do that while we grow our food."

The old man was nevertheless curious. "Have you made any decision about America yet?"

Walton laughed "That sounds very grand and Godlike. My prime responsibility is to hand to my employers the results they hired me to get. That's the simple answer."

"But perhaps not the best one. Is that what you're thinking," asked the old man.

Walton nodded "Yes. I know if I go back and say, *'Hey folks I've got the answer'* they'll just replant all the pines and maybe more and in a few years they'll be back in their old ways, throwing it all away, being as greedy and selfish as they ever were."

"And you feel the decision is yours, not anybody else's. Is that right?"

"I have a feeling that it should be, or that at least the scientist should bear responsibility for his actions, for his discoveries if you like. If I hand over the solution, in many ways, I'm opting out of all responsibility. I'm, in a way, giving them a licence to repeat all their old mistakes. If I don't hand it over, at least, I'd have saved them from that."

"And you feel that once you hand it over then the control is out of your hands and if things went wrong you'd feel guilty about it."

"Something like that."

"But if you don't hand it over you are presuming to make a decision concerning the future of a whole nation. Was that what you were asked to do? Isn't that as you said just now -- rather grand and Godlike?"

"I suppose it is."

"So, doesn't it then come down to a matter of what trust you have in your country to make the best of what you can now give them? Isn't the question simply, *'Will they have learned any lessons from this episode'*? You obviously suspect that no lessons will have been learned, is that right?"

Walton agreed that was so.

"In those circumstances you have to decide whether you have the right to withhold the information. Once you tell anybody else that responsibility is gone from you. It is something you have to decide alone."

"Very much so" agreed Walton.

Quietly, the old man observed "A lot of men would find that a reward of ten million dollars, a measure of fame and adulation, an assurance of prestige, and all that goes with it, difficult temptations to resist."

Walton smiled "Yes and there's all that too, isn't there?"

The old man rose slowly from his chair. "I think perhaps I have helped you all I can. I have made my decision and I hope it is right. I shall pray that yours will be the right one."

"What would your decision be?" Walton suddenly blurted out.

"My decision" smiled the old man "could not be yours."

***

Walton sat for a while, digesting the discussion. What he had not mentioned to the old man, because it seemed so trivial, were the other factors. Like Grace for one.

Increasingly now he was certain, if he were to fly back to the States, grab the reward and all that went with it, he could, given time, make it with Grace. And, perhaps, this time, for good. Over and over again, working out in the jungle, he'd gone over the options. Lestari meant Nani, and that was precious, and she was wonderful too. He could hardly expect a lot more than beauty, love, devotion -- and Nani too. But he also knew himself. Too well he knew how he could end up getting restless, bored and maybe even causing her pain.

Now with Grace there was no chance of that -- or at least less of a chance. With her there was a shared culture, she had the will to bind him to her and there was no way he could drift off into his old complacency.

Suddenly he made up his mind. He'd go back to Washington and explain it all to her, that he'd finally decided he wanted to do the whole bit, even marriage maybe. Once she knew that, then he'd tell her about the solution -- telling her even before the President of the United States -- then he had to be in with a chance.

So, he thought, that was it. He'd finally made his decision. Back to Washington, recapture Grace and then collect the reward and all that went with it. He grinned to himself, he couldn't see himself playing God. This way was much more fun.

Lestari came to the door. She'd seen her father and he'd told her they'd talked. *Had he made a decision?*

"No" he said "I can't make it here -- that can only be done in Washington. I have to see what the situation there is really like for myself."

"I understand" she said. Walton doubted that she did, but he didn't elaborate. She sat on the floor, leaning on his knee looking up at him. Walton was very tempted.

"And you go in the morning" she said.

He nodded and asked her a favour "Do you mind not coming to the airport and saying goodbye to Nani for me?"

She looked puzzled. "Goodbye?" Walton laughed it off "Well you know what I mean. Airport scenes and all that and I never get to kiss you goodbye properly anyway."

She understood perfectly, rose, kissed him quietly on the cheek and left. She'd said nothing. For a moment he felt lost. Odd that. Normally they said goodbye in bed.

He went back to the room in the cottage he now used as an office. He poured himself a drink and sat looking at the phone. Grace would be in the office by now. At least Powell wouldn't be there. What did he have to lose?

"Mr. Walton's office."

"Well that's something -- at least I still have one."

"Dick?"

"Hi honey, how are you?"

She sounded flustered, phone calls from him had become a rarity, let alone ones sounding as friendly as this. "I'm fine -- how are you?"

"Bitten all over -- and I mean all over. I've been out in the field for a few days and I tell you every single one of those bugs remembered where it hurt most."

"But you're OK apart from that?"

"Yeah. What's happening there?"

She paused "Everything. You maybe heard we have new dollar bills now -- from the new papyrus; and General Motors brought out an electric car, a big one so people don't feel deprived. And everybody's turning country boy and heading out west. Washington's almost empty apart from the Government folk."

"Honey. I rang to say I'm coming home. And I need to see you."

"I'll be in the office sir, just like always."

"No Grace I meant alone."

She had no doubts about what he meant. It was obvious from the tone of his voice. She cursed mentally that it should still bother her." She stalled "That could be sort of awkward Dick. You know?"

"You mean Bob."

"Well ..." she hesitated again.

"Honey. I can't really complain about that, but I really do have to see you on your own. It's important. No, it's more than that. It's the most important thing I've ever had to say to anyone."

"Dick" she chided "you had lots of chances to say important things before. Lots of times you could have said things."

"But that was before honey."

"Before what?"

"Well I can't explain now, just take it from me there are some things that have changed, things that have happened."

"Dick could things really have changed that much? Didn't you once say that it was you that would have to change?"

"Honey, I'm not asking for the world right off. All I want is to see you on your own so that I can explain. Don't we have that much left?"

"Sure, and more." She hadn't meant to say it, but she had, and it was still true. He still had something that churned her up. Maybe that was why she didn't even want to see him again. Maybe it was herself she couldn't trust.

Dick was silent for a moment, the last remark ringing in his ears like victory bells. He knew there was something left.

"Honey you know I always said I'd never use those three big words unless I meant them. Well I think I've grown up enough now to face them. Do you know what I'm trying to say?"

"Dick you clown -- why did you have to wait this long? What were you scared of?"

"That I'd be wrong and that you'd throw them back in my face. It happened to me once."

"Dick, honey. I just don't know. What do I say? You know how I felt and maybe there is a lot of that left. But Dick don't you understand a girl can't go chasing dreams and waiting for them to come true. Now there's Bob, and Dick, he's a nice guy. He really is. He's straight, he's simple, he loves me and he's around. A girl can put up with a lot of that. There's lots that don't get the chance."

"Honey, I can't explain right now, but there's something else I have to talk to you about. But again, it has to be alone. Can you understand what I'm getting at?"

"Dick, I can't work out riddles."

"Honey. Look I'm going to come back there. The next plane. I'll come straight to your place. I guess it'll be early morning. Can you try to be alone?"

"Wouldn't it be easier in the office?" she asked.

"Grace darling. I can't explain. I really can't. Just try to be alone will you? Is there a chance?"

"Dick Walton why do you have to keep barging into my life? I hate you for that you know. But yes, there is a chance. I'll try to be alone. All I can promise is that I'll try but I can't throw somebody out of my life just because you phone me from the other side of the world. You understand?"

"Honey you know I do -- and thanks. I'll see you as soon as I can and hey -- I don't want it to be known I'm in town OK. That's important too. I'll explain it later."

He put down the phone and poured himself another drink. He knew he'd have trouble sleeping that night. He was right. Now he'd made the call he was certain it was the right decision. Just hearing her voice again had brought it all back. Oh, Jesus he'd missed her. It wasn't until now he realised just how much.

Just those words alone *'Sure and more'* made him certain he still had a chance of getting her back. All he wanted to do now was hold her, beg her forgiveness and ask her to come back.

He poured himself another drink. It was 2 a.m. He turned out the light in case Lestari thought it was an invitation. She was lovely and warm and tender and there was Nani, but Grace still had that certain lure and ability he'd never found in any other woman.

She excited him and she was someone he felt he had to live up to. That was the difference, he decided. That was it. Now he just knew he was right. He wondered in passing how he could have been so wrong before.

He thought of Powell, but even that somehow didn't matter. When he got back, when they were together she'd be his then, nobody else's.

He was sure of something else too. Having the solution, then the reward and everything that went with it would make the difference. It stood to reason -- no girl who even half liked a guy would turn him down if he had ten million dollars. But for her, and for him for that matter, it wouldn't be the money that was important. It would be the prestige. She'd have something to be proud of. It would be her guy that had done it all, turned America round. The more he thought of it, that would be what made the difference.

He emptied the bottle. The timing. The timing would be important too. Yes, that was crucial. He grinned to himself, carefully took the bottle by the neck and lobbed it with satisfying accuracy into the waste basket. He'd cracked it now.

He'd tell her first of all. Then take her with him so that she could be there when he told the President. How about that? It was a master stroke. How could she ever forget a moment like that? Why it was history in the making. There was no way a girl could turn down that sort of invitation.

He could see it all, the President, the Oval office and Grace by his side. He'd have to think of some great phrase, something that would stick in the mind, some remark that would become part of American folklore.

"Mr. President I give you a tree."

No that wasn't it. "Mr. President -- America's back on the road." Maybe Grace would have some ideas. She'd like that. "Mr. President -- I bring you trees."

No, it was no good, he'd think of a better one later. The booze had got to him now. He wondered if there was another bottle. And then, he thought, there'd be all the chat shows. He'd be serious on those, take the chance to tell America how lucky it was to have another chance, impress on them just how important trees were -- how this time they had to be taken care of.

Then there was the money -- oh what a great trip they could have. Right round the world, apart from Indonesia maybe. But Europe certainly. And then, the thought struck him like a thunderbolt. There might even be a Nobel Prize -- or at least a share in one. Now how about that?

He was really going to enjoy all this. He could hardly wait. All he had to do was to see Grace first. But how could she say 'no?'

***

He waited next morning until he heard Lestari's car leave with Nani. He felt guilty about it, but in an odd sort of way even saying goodbye would have been a breach of faith with Grace. And if Nani had asked him when he'd be back he wouldn't have known what to say. He wouldn't have had it in him to lie to her, yet he couldn't have told her the truth. There was a sadness in him then. It lasted until he arrived at the airport. Then the old drug took him over again. The smell of that fuel, the whine of the big jets -- he could feel the old tingle still go down his spine.

He looked around, as if for the last time and remembered it with affection. It had been good here.

In his mind an album of pictures flashed its way past -- Lestari meeting him the first time, and her in bed, and looking so gravely attentive as her father spoke at dinner. And on the beach and holding Nani. The thought still broke him up. He dismissed it instantly, or at least as quickly he could.

He hoped the plane would have some newspapers gathered on the stopover. At least he could bring himself more up to date that way. Voice of America was fine, but it was fairly formal and, more often than not, it meant the voice of Bob Powell and that just made him mad. And all such pious bloody stuff about people learning to live in the country again. Joe Simon though, he had to admit, had made it sound better. He was all for life as it was heading now. He really believed it, but he'd still not be able to turn down the chance of starting from scratch, this time making the forests work for the people and not Wall Street. He would never object to that.

Yes, it was going to be a good time. Oh Grace, oh Grace how you've been missed, he daydreamed. Ye Gods it would be good to see her again. He couldn't wait. But he had to. He had 10,000 miles to wait. Time for a drink. And maybe a real dream.

Ten thousand miles later. The American dream. Nothing had changed, there was just a slight shift in the course, a tack along the

course of time. And he was at the tiller. He Dick Walton had the hand on the tiller of power.
It felt great.

## CHAPTER NINETEEN

Lean was back again at the window of the Oval office -- and he seemed depressed.

Joe Simon had called in to brief him on the morning's meeting of the allocation committee. It was a job he hated, sharing out a decreasing amount of money between an ever-increasing number of demands.

"You know Joe -- I'm the only President who's ever had to run an America that was broke. You think I'll make the history books?"

Joe stayed silent for a while. *'What history books?'* he thought.

Lean seemed to shake himself back from his musing. "So, tell me what happened."

Joe explained that the money for welfare payments was now almost depleted. While the government had to pay out more in welfare cheques its income from taxes -- on cigarettes for example -- had gone down. The hard-liners had resisted any more money going into the welfare budget and had insisted the individual states take over the job of handing out the money. Joe had felt like a Judas again, he knew what would happen to the poor in some of the states. The best deal he'd managed was that the states would dish out the money but use the guidelines from Washington. They could argue the finer points of that later on.

Lean took in the news. Though he'd been elected on a liberal ticket there were some harsh decisions that had to go the other way. For the moment they had no alternative. Lean briefed Joe on the foreign affairs scene.

He went back to his desk. "You know in the old days it was always *'What does America think? What would its attitude be?'* Almost always they were after some sort of handout, some sort of favour. Oh boy have things changed. It really tells you who your friends are, and I tell you Joe we don't have that many."

“We can just thank our cotton socks for Europe. Without them we’d be on our asses.” Joe nodded. He knew the European Union had been sympathetic.

Lean went on “Every little bitty country these days is making us pay for what guys like Daley did to them in the past. And boy can they be petty. There was an embassy party in Tokyo the other day and our guy could only get in as a guest of the Israelis.”

Joe grinned “I guess we still have our uses.”

But now the banter was over, Lean realised Joe had something on his mind. Simon outlined the welfare problem and stressed how careful they must be that the States kept to White House guidelines. "If we don't then in some places we could have a lot of trouble. There's a lot of hungry folk now."

Lean agreed he’d make sure they stuck by the rules. He had a meeting that afternoon with senior congressmen and senators and would remind them once again that it was Daley's fault in the first place. That usually shut them up for a while.

Joe mentioned his scheduled interview with Powell and agreed to the general line with Lean. Then he mentioned his "rent a farm" idea, stressing he had to check out snags with Walton first.

Lean looked thoughtful. "Could it work?"

Joe thought it could; it could also have the additional advantage of mopping up some of the unemployed from the downtown areas.

Lean was doubtful "You know how they can be in Kansas don't you Joe? I mean if you flood 'em out with young blacks from the cities -- isn’t that going to cause trouble?"

Joe had thought about that. "I reckon most farmers are in it for money -- if they weren’t always, they are now. If they'd be getting rent for land that they wouldn't otherwise be using, and they can still grow grain I can't see that they'd object too much. The other point is that the State can fix the rents and can take some out along the way for taxes. It seems worth a try."

Lean nodded. "OK. Let's give it a whirl." He stopped for a second. "Speaking of farming, what about that papyrus stuff they were trying to grow? Did that work?"

Joe smiled "It sure did. Even the big timber companies were in on it. They've planted up whole areas now and in a couple of months they reckon they should get the first cut. But it'll just give us some paper, not much else."

Lean went back to the window and Joe left.

***

Powell was prompt -- and he'd already picked up whispers about the welfare payments decision. Joe smiled "I just love the way one thing never changes in this town. The only thing still working overtime is people's tongues."

Powell laughed. "I'd sure be in a mess if they stopped."

They did the interview, Powell playing devil's advocate on the welfare decision. "How could a government that rode into the White House on a liberal ticket ignore one of its prime duties – to look after the poor? Doesn't it care anymore?"

Joe agreed it was bad; he was sad about it too, but he reminded Powell of the point made in their first interview: that part of the policy now was to have less government not more. The States would administer to their local problems and the guidelines would be strict. And, he added, the White House was thinking of taking the initiative in a new policy designed to help people become self-sufficient.

He explained the outline of his rent a farm scheme, stressing that it had worked elsewhere in the world with regard to forestry. Powell was sharp, and like Lean, he spotted the threat of trouble on the plains.

Joe came back quickly "Mr. Powell, only days ago you filed a piece from Kansas saying it was happening anyway. We're just easing the path a little, making sure people have a chance to earn their own living and recover some human dignity. Part of our function is to ensure they don't get ripped off in the process. Can you argue with that?" He grinned impishly across the recorder at Powell.

The reporter came back "But the farmers are choosing who they want to have -- maybe they won't want the White House dictating they take young blacks from the cities. Maybe they should have a choice. After all, it's their land."

Simon chided him "I had a feeling in the past that farmers thought young blacks -- and old ones too -- were suited for nothing else but picking cotton. I thought we had that problem licked a few years back. And just think, this time they don't even have to pay them. They even get rent -- and they can still pick what land to let."

Powell still came back "But that whole agricultural system is based on vast plains, huge machines. That's the only way it works. You sure it's practical?"

Joe had an answer for that too. "There is an argument for *small is beautiful.* At the moment all that land is used to make a lot of money for a small number of people. This way the benefits get spread around a little. And maybe, Mr. Powell, since you come from the town, you don't know too much about the dangers of having huge areas devoted to one crop year after year. It can only last so long. There is a very practical argument for going back to old fashioned methods and letting the land rotate in the old way to retain its goodness. Maybe you didn't know that?"

Powell was caught. From the research he'd done while in Kansas he knew Joe could be right. He changed the subject but kept up the attack "Sir, the impression round the country seems to be that Washington is out of touch; that it doesn't realise the way the crisis is hitting folks in the small towns. What impressions do you get from your vantage point here, about the rest of the country?"

Joe thought for a second, then explained. "Firstly, you can't sit in this office and not know about the pain. Remember I'm from Oregon. How do you think I feel when my folks tell me it doesn't have trees anymore? How do you think I feel when I have to sit down and make a choice between resources for Defense and Education? How do you think I feel when I hear of an old rabbi who died because somebody stole his scroll? No, I guess I know enough about the pain that's going on. We have some here too."

He paused again, Powell looked up anxiously, wondering whether to ask another question. Joe held up a finger and continued.

"But let me tell you something else that's going on -- and maybe you sensed some of it too. You know I sometimes get the idea that although it's hard right now, we may just be learning a few good lessons along the way. There are a lot of people helping each other more now than they used to. We all know this country has people from all over the world and they stick together. They help each other when there's a need. Same goes for people living on the same block. In all sorts of small ways there's a new sort of feeling around this place. Let's just say folks do move back out into the country. They'll have to fend for themselves and country folks have always had to be good neighbours.

And there's more folks in church too. One of the things I never liked about America too much in the past was its apparent greed and selfishness. Now I think a little bit of that has disappeared. I know it's hard going shopping -- I do some myself. And I know it's kinda hard not to ride around in the car so much. But people are fitter now and a lot of the old American know-how is coming back. I don't think it's all doom and gloom. I still have a lot of hope left."

Powell switched off the machine. And was silent for a while. Joe looked across anxiously. "Was that OK?"

Powell rewound the tape and looked up. "Yes, that was more than OK; it was great. What's more you could even be right -- about folks pulling together and all that." He smiled "I'm still not sure about cotton picking blacks, but who knows?"

Joe apologised and laughed slightly. "I'm sorry about that, but you know how I feel about discrimination. I even feel guilty we call this place the White House -- what sort of name is that for a country with some many mixed races?

Powell couldn't resist a gentle probe. "You any nearer a solution? Do they have something to finish off that PWN stuff?"

Joe shook his head. "We have an awful lot of people working on it -- that guy Walton and the rest, but I don't think it's easy."

"Is he still here?" asked Powell.

Joe didn't know. "Last I heard he was head down in a library somewhere. I guess you should ask his secretary. You want me to call her?" Powell said not to bother. He'd drop by on the way out. He thanked him again for the interview and Joe in turn congratulated him on the pieces from Kansas and Detroit once again.

They had been good and, incidentally, Walton's secretary had liked them too. "She seems to be a fan of yours." Simon noted.

Powell smiled "The feeling's mutual."

## CHAPTER TWENTY

*'Don't we have that much left.'* Walton had asked when he called Grace.

*'Sure, and more'* was her answer.

The words had sustained him through the long flight. He checked his watch, setting it to Eastern time, and went to change his money. The dollar bills were new, rougher now that they came from the new papyrus mixture. Walton smiled to himself, trust America to make new dollar bills a priority. He re-checked his watch. Another day, another dollar. Or maybe ten million?

He wondered about taking flowers, then discarded the idea. The traffic that early in the morning was almost non-existent but looking at the gas prices -- seven dollars and fifty cents a gallon; he was hardly surprised.

It was 7.30 when he arrived at the flat -- maybe she'd have some coffee going. She used to bring it to him in bed, never bothering to dress. And often it had gone cold. Her car was there, a compact bearing the battle scars of downtown traffic. And the light was on in the flat.

Outside the door he paused, took a deep breath and knocked. He saw the spyhole cover go back, followed by the chain going on and the door being unbolted. Was crime really that bad now? His smile was already starting, an intimate friendly grin.

A man opened the door and looked at him rather blankly. Walton checked the number on the door. It was Grace's flat. So, this must be Powell.

"Are you Dick Walton?"

"I came to see Grace."

"She said you were coming. I'll get her."

"I could come in you know -- I've been here before."

Powell looked embarrassed for a second. " I'm sorry. It's early yet."

Walton felt sick as he shambled into the kitchen. Grace, in a housecoat, was attending to the toast. Powell stayed in the doorway. Walton felt awkward, like he was on display. He wondered when Powell would leave. Grace turned, slightly flushed in the face, "I'm sorry. I tried but..."

He relaxed slightly. Well at least she'd tried to get rid of him and obviously he wasn't going to leave them alone. But at least she'd tried. Maybe he could help her. "Grace I need to see you alone." He said it loudly enough so that Powell couldn't be mistaken. Powell coughed quietly. "Listen, Grace and I talked this whole thing over last night." It was Grace who looked embarrassed now and she turned quickly back to the toast. It seemed to be demanding a lot of attention, mused Walton.

Powell continued. "I've asked Grace to marry me." Walton felt as if he'd been kicked in the guts. He put a hand to her shoulder, whispering almost "Grace there's something I can only explain to you on our own."

She turned then, her mouth just taking on a slight firmness.

"Dick, Bob and I had a long chat last night. I know we have work secrets from each other -- that has to be -- but everything else is out in the open. So, Bob has a right to be here. And I'm sure it isn't about work. We can always talk about that in the office."

Walton's heart sank. The instinctive reply that it was about work, well partly anyway, died in his throat. Her phrase about secrets rammed it home again that Powell was a reporter. And cute too. There was no way he could even drop the smallest hint that he had a solution.

She went on "If it is work of course I'll come in, but you wouldn't be here if it was and I was planning to have the day off today. I didn't have a chance to ask you, but I've found a stand in. She's very good and she can handle anything in the office."

Walton didn't know what to say. "On the phone" he stumbled. "On the phone I thought there might be a chance..."

She didn't dare look at him. Seeing him here again now still left something happening to her stomach. The tan suited him and there was an odd grey hair now. The distinguished scientist look would suit him.

Powell coughed in the doorway. "Grace honey, you were going to give me a lift. You going to get ready?

"Yes, I won't be long."

Instinctively Walton made to follow her to the bedroom.

Powell didn't exactly stop him, but just shuffled his feet.

"Dick. I believe she has made up her mind. At least I hope she has." He had a slight nervous cough. "I guess I know I'm sort of second best and that if you'd still been around I wouldn't have had a look in, but we do get on well, and...

"And you couldn't wait to jump into that bed the second I was gone." Walton felt angry now, he ought to punch this guy. He pushed past him. Grace had already raced into the bathroom. "Grace all I need to do is to see you on your own so I can explain. Surely you can trust each other enough for that can't you." He turned to Powell. "Can't you?"

Powell apologised. "I said I'd be here. Grace wanted it that way."

Walton was certain he was lying. He had to be. She'd even said she was sorry Powell was here. Could she have been lying? If only he'd had some paper. A crazy thought but he could have scribbled a note and slipped it under the door to her. This was no way of going about things. He needed to be able to look at her, he knew how to do that, what phrases, what sort of voice, what sort of face to use.

He turned down the hall, feeling ridiculous. If only he could tell her. He swung on Powell, still angry with him. "You talked about marriage. What did she say to that?"

Powell tried hard not to look pleased, it could so easily translate into a gloat. "She reckoned I stood a chance."

"Yeah I thought so too, once" said Walton bitterly.

Powell coughed again. "The way I heard it from her you didn't exactly make it a firm offer."

"But I would have done -- it was just a matter of getting around to the idea, waiting for the right time."

Grace left the bathroom and in a minute was with them, ready for the outside world. She slipped her hand into Powell's. Walton winced. He was so angry now, so frustrated at the illogicality of the whole situation he was almost shouting now "I've come ten thousand bloody miles. Surely I get to have two minutes with you?"

Hesitatingly she looked up at Powell. He half shook his head. She looked back at Dick, not finding it easy. "Dick, no. We talked about it."

She added "What I'm going to do today is to try to find another job. Maybe a foreign embassy here or something. It would be easier for Bob that way. You know the way people talk. How bad would it be if I didn't come back to the office?"

He had to give her some clue, just some small hint. "Surely -- there's nothing to do now…"

"What do you mean" she asked "Nothing to do, surely...?"

Walton said quietly. "It's finished." Grace took it the wrong way, presuming it a reference to their affair. "Dick we've been through all that. Look we have to go now. Bob's has to get to the office. I've got the car and well...."

Walton shook his head again. "Don't bother. I'll call a cab. I have to see someone too." Suddenly, impulsively she rushed forward, pecked him quickly on the cheek and had moved away before he could react. The door shut and they all walked down the stairs together. "See you fella" said Powell.

"Good luck" said Walton forcing a smile.

"Dick, I'll see you" waved Grace.

Just for a second she paused. "Dick are you going back to Jakarta? Or are you going to Portland? It's just the office will need to know where you are."

In a moment of blinding temptation, he almost blurted out the truth. Powell got out of the car again. The sight of him riled Dick now. He was sure he was using Grace. He had to be getting information from her. Just one word from him in the wrong place now and he'd have it on the air in an hour, despite any pact he had with Grace. He knew those reporters, especially the ones like Powell that hung around the bars on Capitol Hill.

He very slowly shook his head. "I'll deal with the office. Don't you worry about that."

Just momentarily she paused, then Powell muttered something to her, and she got back in the car. The gears crashed, she shot headlong into the traffic from the drive and Walton went back to the apartment block.

He found the pay phone, noticed the lack of a directory and called a taxi number chalked on the wall.

"Yes, that's what I said -- the White House."

***

He slumped against the wall outside to wait, a bitter feeling inside him now. She'd lied to him about not wanting Bob to be there. He was sure of that. And she'd always wanted to be married too. Powell in his own way had the power that attracted her. Not ten million dollars of power, but he hadn't been able to even hint at that. Maybe he had just left it too long this time.

A girl in high heels swung her hips passed him. She was trim. She paused for a second. "You are waiting for someone?" He smiled at her. "I could do with a lift if you're going downtown."

"I sure am, honey. You just come along with me Where's it you're going to?"

"Would you believe the White House?"

She nodded. "Sure do-- there's some folks here in this block who work there. Maybe you know them?"

All the way downtown she talked, intrigued when he said the tan was acquired in Indonesia.

"Say isn't that the other place that lost all its trees?" He nodded that it was. " I'm going to be so glad when we get those back. You wouldn't believe what a girl has to go through these days. I mean the bathroom just isn't what it was and as for clothes - you see the price of those now? Shopping is another nightmare altogether, taking all those containers and things to carry what you buy."

"But you can still get food OK?" asked Walton quietly.

"Oh sure, of course we can. Well most things anyway. I don't know whether you know, being away from the States, but we had to sell some food overseas to get money in for other things. That meant we didn't have any buns for hamburgers and there were no napkins either -- I tell you it makes things real messy. And that rosewater stuff they have in restaurants now, it can really dry up a girl's skin. Here we are now. Right outside your destination. You can just walk in from here. Any time you want a lift you just call. Ginny's the name. Apartment 17."

Walton smiled at her. "Thanks, Ginny. I might just take you up on that."

***

She drove off, leaving Walton on the sidewalk. It would have been easy to put her right, but what was the point? If she didn't know by now, she never would. Jesus, what a bloody selfish nation it still was.

He fished out his pass and headed along the corridor to his office. A new girl looked up, asked who he was and looked flustered.

"I'm afraid Grace isn't here today. She..." Walton smiled at her. "I know. I've just seen her. Don't worry."

"Can I get you anything, sir?"

"Call me Dick and the answer's 'yes'. I know it's early by the watch, but in my head it's the middle of the night sometime. How about a coffee and a bourbon?"

"A bourbon sir?" He stood up, went over to his own desk and withdrew a bottle.

"You bring a glass and a coffee. I'll do the rest."

She bustled out of the office. Bourbon in the White House at nine in the morning. Wait 'til she told the folks. She came back with the coffee, he sipped some then topped it up with the spirit.

She looked at him expectantly. "Anything else sir? Dick?"

He came out of his reverie. "Yeah. See if Joe Simon's in. I'd like to talk to him as soon as possible. It's important."

"The Vice President?"

Walton sighed. "Yeah. His name's Joe Simon."

She looked flustered again and a bit snubbed.

"Honey I'm sorry. It's been a long day. Just see if he's there."

She busied herself with the phone call to Simon's secretary, taking the number from others scribbled on the wall. Seconds later she looked up "He says to go straight in."

The bourbon had hit Walton's stomach now, lifting him slightly. He stiffened his shoulders as he walked along the corridor and into Joe's office. Joe bounced up from an easy chair; there was no point now in him having a desk. "Dick -- great to see you again. How are you?"

Walton returned the handshake and went to the window. "Joe do you think we could talk outside. Say on the lawn? I could use some fresh air."

"Why not. Nobody ever uses that lawn. Seems a pity. Now tell me what's been happening." They went out through the French doors on to the lawn. Dick noticed a small pine sapling in the mid distance. Joe explained "We got some from Korea. Had a national tree day. The President planted that one himself -- right outside his office. Didn't you hear about it?"

"I guess I missed that one" confessed Walton. "I've been out of touch a bit lately." He turned to Joe. "Tell me what else has been going on. I hear you don't have buns for hamburgers anymore."

Joe laughed. "I guess that did more to bring it home to people than almost anything, bar the lack of toilet rolls. And newspapers maybe. Funny how people miss the trivial things."

Almost like a roll-call he filled him in on the developments of the past months. There was the share a farm scheme, now proving a big success and that was giving them the chance to do something about the inner cities. Joe was bubbling at the thought of making places like Detroit resemble Oregon. There was the electric car, new dollar bills, people generally were healthier, but not just in body.

"I'd never have believed it in the old days, but you can almost feel the energy here these days. It's somehow given people a challenge and with something big like this around a lot of the old petty grumbles have gone. It sounds crazy, but do you know at the latest count three quarters of the people turned up to pay their taxes -- they'd kept hold of their money and paid it in as good as gold. They reckon it has something to do with the way religion's asserted itself again all over the country. But it really shook some people here. Even my own rabbi was saying we had to set an example… In the old days he was the guy who'd claim his robes were deductible."

Walton listened carefully. He'd heard bits of it now and then on Powell's broadcasts on the radio. Hearing it all in one fell swoop made it a huge catalogue of change. "But you still must have problems surely? What about the climate and erosion?"

Joe agreed. "But even on the climate question, it hasn't been that bad yet. It's been hotter and yes the crops weren't too special first time around but having less cars on the road has helped. All the dams have been dredged right out we've been able to cut back on using coal for electricity. That's helped a lot. About the erosion thing, there's a guy making a fortune with some silly little plastic things that help build up terraces to hold the water. It sounds crazy, but it seems to work. There's lots of new stuff being planted too in the way of crops, not just corn so it gets a rest but different vegetables and things like that. I reckon by next year we should have around ten per cent of the paper we need just from papyrus. Also, there's still people experimenting with bagasse. It's not all bad"

He paused "In world terms we're poor and the dollar is still low, but that does mean it is much easier for us to sell anything we export. The solar panels are a good example. We can make them and sell them cheaper than anybody now. People have seen what happened here and they're taking notice. I think what it comes down to is this; by and large, people now have what they need, rather than what they want. If anything, you know that's been the most impressive thing of all - the way society has broken itself down into smaller units. The government now is in the State capitals hands far more than here, and that's done them a lot of good too. They've always known what's best for their own people."

Walton stopped him for a moment. "You almost sound, Joe, as if you like things better this way."

Joe laughed out loud. "John's always telling me that. And Ruth too. That's why I always have to help her with the shopping now, just so I can be brought back to earth. But seriously I'm not sure we aren't better off this way."

Walton phrased his next question carefully. "So, what would you do if someone came along and said, *'OK I've cracked it'* I've got the solution to PWN. We can go back to the old ways.'"?

Joe looked at him quickly for a second. "I guess the President would clap him on the back, hand him ten million dollars -- on TV of course -- and make him a national hero. And I guess he'd think with the mid-term elections coming up it couldn't happen at a better time. There are still greedy people around, even in all this."

"And what would you do?" asked Walton quietly.

Joe looked at him again. "You know I think my first temptation would be to ask the guy to go away and think about it. I mean it's a hypothetical situation we're talking about isn't it?"

He glanced quickly at Walton again. Walton stared straight ahead. Joe continued "I guess the guy would be very tempted by all the money and the rest -- no problems with girls for the rest of his life."

Walton glanced at him sideways but said nothing. Joe went on "I guess that would be important to some men, but overall -- in all sorts of different ways -- I'd be sad to see all this new way of doing things and just plain living get dropped for what we used to have. A lot of it's been worth getting back into, our way of life as it is now." He turned to Walton to look at his face for a second.

"Talking of girls and it's none of my business -- but Ruth's bound to ask you know -- is it true about you and Grace...that it's not on any more?"

Walton agreed silently, the merest nod. He explained "She's with Bob Powell."

Joe said he'd heard; she'd always liked to be at the centre of things hadn't she? And, she had the brains for it.

Walton said that was right. He shrugged his shoulders. "I guess the way she is now she wouldn't be tempted back by a million dollars."

"Or even ten?" asked Joe.

"Even that" agreed Walton.

Joe said nothing for a while, then turned to Walton again. "The guy who found the cure -- saying it happened -- he'd have to have thought a lot about his responsibility in all sorts of ways. I mean not just to America, but Indonesia and even the world at large."

"He would" agreed Walton "though I didn't tell you they've found a way round it. They're going to play it cool, grow some food, plant some trees and use that time to learn how to process their own logs. Then they'll just sell to the highest bidder. So, I guess they'll be OK."

"So, the guy wouldn't have to consider them would he?" asked Joe.

Walton shook his head. "Not at all. It would just be America he'd have to decide about. God-like sort of decision isn't it; wouldn't it be?"

Joe smiled at the slip. "God's decision made by a man. Tough. You know there's a lot of people -- and I guess the President would be one who'd say any guy who was hired to find a solution owed it to his bosses to give it to them."

"If the guy worked for them..."

"Yes. If." said Joe.

"And, as you said, I guess the guy would find it hard to turn down ten million dollars and the rest that went with it."

"What it comes down to" said Joe "is it would have to be up to that man to make up his own mind. He'd have to look at America now, decide whether it was better than in the past, then make up his mind."

"What if the guy came to you for advice Joe – or what if you had to make the decision for him?"

Joe paused in his walk, looked directly at Dick and said "I'd tell him I'd rather have things the way they are now. I'd say it has every chance of being a great country all over again – but, in the proper way this time. That's what I'm saying."

"Or would say if the guy asked you?" said Walton.

"If this guy asked me, that's what I'd tell him."

"And he'd get his reward in heaven."

"He'd have to wouldn't he. On the other hand, someone in government would realise that the guy was pretty bright and I'm sure they'd find him something really important to do. I don't think he'd starve."

"I guess he might be OK at that" said Walton.

"I'd make sure of it if I were in government and it happened" said Joe, not looking at him at all now.

"Dick" he continued. "I think when it comes down to it. When it really comes down to it I'd tell him I didn't have any right to ask him to turn down ten million dollars and all that went with it, but that I'd not want it announced. I'd want us to carry on in the new way. Because I think it's better. But, at the end of all that, I'd have to turn around to the guy and say 'Buddy it's up to you. And I'm glad I don't have to make the decision. Only you can make it.' That's what I'd tell him."

They were facing each other now. Walton leaned across and shook his hand.

"Thanks Joe".

"Thanks for what?" grinned Joe.

Walton turned away. "See you." He walked slowly across the lawn, pausing to pat the sapling as he passed the President's window.

Joe went back to his office.

***

Lean and Joe were having an evening drink in the President's office, going over the events of the day.

"Did I see you with Walton on the lawn this morning" asked Lean. "What was that all about?"

Joe shrugged it off.

"Nothing really. I think he's got some personal problem. Gotta make a decision I guess."

Lean's interest lapsed. Then a thought struck him.

"Say. Did he bring me any cigarettes? He said he would. Shit. I bet he forgot."

Joe smiled wistfully. "Guess he must have. Must have had something else on his mind."

*The End*

**AUTHOR PROFILE:**

Graham Mole, a Journalist/TV producer with a host of awards. started out in the workforce as a Junior reporter on a local weekly paper, then freelanced on news for national papers before switching into national television, firstly as a researcher, then as a producer and director. He has also been a ghost writer for (David and Charles Publishing House, UK)

He lives in Eastleigh, Hampshire UK on the edge of a forest created by King John for hunting and is an ardent defender of the environment.

He is as a specialist in 'green' & conservation issues and currently writes feature articles for the national magazine *"Forestry Journal"*.